BUILDING VISIONARIES
THE UNSUNG HEROES

Caroline Collier

Published by The Chartered Institute of Building

Contents

Produced 2009 by

The Chartered Institute of Building

Englemere, Kings Ride, Ascot

Berkshire SL5 7TB, United Kingdom

t: +44 (0)1344 630 700 **f:** +44 (0)1344 630 777
e: reception@ciob.org.uk **w:** www.ciob.org.uk

Registered Charity 280795

Project managed by Saul Townsend.
Designed by Stephanie Lee.
Photography by Matt Wain Photography.

Set in Minion Pro Roman.
Printed in Great Britain at Alpine Press Limited.
ISBN 978–1–85380–499–1

Acknowledgments

I have been grateful for the help of many people during the researching and writing of this book. Firstly, every living person featured made themselves available for interview, and their generosity with their time is greatly appreciated.

Many staff from the Chartered Institute of Building (CIOB) assisted, and I must particularly thank Liu Mengjiao, who interviewed Professor Xu and transcribed their conversation into English – an invaluable contribution for which I am most grateful. I'd also like to thank Rod Sweet, editor of iCON, who was also extremely generous with his notes from interviews with Professor Li Shirong and Yi Jun. I am greatly indebted to his kindness.

Friends and family members of some of the leaders mentioned also made an invaluable contribution to the book. There were so many interesting and insightful contributions that I fear that there are some I may have overlooked – many thanks to them all, including Lady Valerie Dixon, Steven Dixon, Peter Harper, Dr. Hew Hioen On, Sir Brian Hill, Mike Johnson, Terry Kenny, Sir Martin Laing, Christopher Laing, Ron Malyon, Paul Shepherd, David Trench and Margaret Valentine.

Finally, especial thanks to Professor John Bale for his extensive contribution of information on the development of the Institute, and also to Professor Roger Flanagan, and to the CIOB's indefatigable past CEO, Dennis Neale OBE, whose hospitality and generosity with time and information could not have been bettered.

Caroline Collier
Author

April 2009

Millennium Dome | London, United Kingdom

Foreword

There's a perception that building is all about bricks and mortar, but in reality it's all about people. We rely on our built environment, not only to satisfy our basic needs for safety and shelter, but also to deliver an increasingly complex range of specialist functions. From hospitals to technology parks to schools, we build for a stakeholder group which encompasses every member of society. In essence, we build so that people can have a better quality of life.

People are also at the heart of the construction process. Less than half the value of a building comes from its material components – the rest comes from the skill and dedication of the people who built it. The construction team will always need the contribution of a range of specialists – from architects and engineers to facilities managers – but the most vital component in its success is the vision and inspiration of its leader.

In 1834, a group of the most successful building contractors in London combined to discuss matters of mutual interest. Soon, they were known as the Builder's Society. They could not have foreseen their society's gradual development into the Chartered Institute of Building (CIOB), now a global organisation which provides a professional home for specialists from every part of the building process, but the great strand of continuity between then and now is the leadership the Chartered Builders provide to the dynamic and multi-faceted entity which is the building industry.

This book, commissioned to mark our 175th anniversary, commences with a short history of the Institute, from those early meetings in a London coffee house to the international organisation of today. But the real story is about the people – the incredible courage and leadership shown by the people charged with the job of building the landscape which shapes our lives.

Many of the leaders featured here led major projects as construction managers, but the CIOB is a broad church, and our roll of members also includes some of the most talented of our fellow construction professionals. The engineer Sir Ove Arup was an honorary fellow, whilst architect Sir Edwin Lutyens was the founding father of the ASI – now our Faculty of Architecture and Surveying. Both leaders are featured in the pages that follow.

It's good to celebrate our history – it's a fascinating account of the development of our profession. But perhaps the most important chapters of our story are yet to be written. Chartered Builders have never had such a vital role to play – their response to the current challenges of future planning and sustainable development will be crucial to the well-being of all future generations.

Meeting the need for change in the way we create our built environment will take the work and commitment of our entire profession. It will not be easy. But there is no better place to draw inspiration for the task than from the achievements of builders through the generations. As the following pages show, time and time again they have produced results in the most demanding situations. There could be no more resourceful and committed group of professionals to steer us to a more sustainable future.

Chris Blythe
Chief Executive of the Chartered Institute of Building
April 2009

Broadgate | London, United Kingdom

Introduction

By creating many of the world's iconic structures, today's Chartered Builders continue a tradition of innovation, ingenuity and engineering prowess. Indeed, the legacy of CIOB members includes landmarks such as Nelson's Column, the Houses of Parliament, the Petronas Towers, Sydney Opera House, the Millennium Dome and the Beijing Olympic Birds Nest stadium.

These globally-recognised cityscape highlights are the realisation of some of society's most ambitious and inspirational building projects. But, while those involved in the creation of these striking edifices should be proud of their achievements, the contribution of construction professionals to society is not always widely recognised. Moreover, the path to recognised status for the Institute representing these professionals has been long and often tortuous.

The foundations

It is believed that the first meeting of the Builders' Society came about because of an invitation to tender for a contract lacking an arbitration clause. However, industrial unrest soon became the focus of the members' concerns: in 1834, there was a major strike.

In the spring and early summer of that year, the master builders had met regularly to review the rising levels of union activity. The numbers in attendance would have been far greater than the official membership of the Builders' Society because meetings were publicly advertised.[1] In June, the masters convened at a coffee house on Ludgate Hill and passed a series of resolutions – including one that required their workers to renounce union membership.

The master builders tried to justify this position in their Statement of the Master Builders of the Metropolis. To a modern reader, it is evocative of caricatures depicting the union excesses of the 1970s. The men would strike if they didn't like the foreman, if they were requested to perform overtime, or if they were asked by name to do something (instead of being left to decide for themselves who would comply with a general instruction). The statement concludes that "The natural consequences of these instructions soon appeared in wanton and dictatorial interference with the master builders of the metropolis in every department of their business… and all this must be submitted to, or the work would be suspended, and the penalties attending its non-completion incurred."[2]

The workers were predictably unimpressed. The Central Committee of the Operative Builders issued a reply that challenged the veracity of many of the masters' claims, saying that the Statement was full of "groundless and iniquitous" charges. They also entirely refuted the masters' contention that signing away their union rights could ever be construed as a conciliatory or middle course."[3]

The masters' assertion that they hadn't formed an association themselves failed to convince the workmen, who pointedly added this postscript to their statement: "Employers not connected with the London Coffee-house Union, requiring steady Workmen of first-rate abilities, may obtain them by applying at the Operative Builders' Offices, 12, Little Portland Street…"[4] The situation was at a stalemate.

It wasn't until November 1834 that the strike was officially over, with both sides claiming partial victory. The workers had permission to form another union; the masters saw the demise of the Operative Builders Union only a few weeks later, and had their men back at work. However, both sides had suffered a loss of earnings, and neither had gained exactly what they wanted.

The cornerstones

While the London Coffee-house Union (as the workmen disparagingly called it) quickly disbanded, the Builders' Society proved far more enduring. It's not possible to identify all 17 founder members, since the Institute's first book of minutes (covering the years 1834–1859) has long since disappeared. However, we do know that in 1848, the Society included Elger, Grimsdell, Grissell, Haward, Herbert, Kelk, Lee, Peto, Piper, Rigby, Smith and Stevens – many of whom would have been founder members. If we

1   Holt, T. 2004. 'Masters of their own throats': the 1834 beer strikes Brew. Hist., 114, pp. 22–33 Available at www.breweryhistory.com/journal/archive/114/bh-114-022.html Accessed 26th August 2008

2   Evans et al. 1834. Statement of the Master builders of the Metropolis relative to the differences between them and their workmen on the subject of the Trade Union. London: Dean and Munday p. 4
3   Styles, JD. 1834 Reply to the Statement of the Master Builders. p. 6
4   Ibid. p. 8

add in the names of Thomas Cubitt and his brother William, we have a total of fourteen. Others, such as Baker, Lucas and Myers, are also known to have been early members.

The Cubitts are studied in more detail on page 12 and Grissell is discussed as a business partner of Peto (see page 22), but what of the others? Among the most distinguished were Kelk (who built the Albert Memorial designed by Sir George Gilbert Scott) and Myers, who was known as Pugin's Builder because he carried out many commissions for the great architect, including his home in Ramsgate.

Other early members were equally illustrious. The Lucas Brothers, who learned their trade from Samuel Morton-Peto himself, built the famous Oxford Museum and the Royal Opera House in Covent Garden. Grimsdell was involved in rebuilding the Fishmongers Hall and creating a library for London University, although perhaps his most notable work was elsewhere: he helped build dwellings for both the Society for Improving the Condition of the Labouring Classes and for the Metropolitan Association for Improving the Dwellings of the Industrious Classes.5

Finally, an overview of the achievements of the early members would not be complete without a mention of Society founder Henry Lee. His family contracted for most of the trades work on the north wing of the British Museum and subsequently won the first part of the building contract for the Palace of Westminster (normally remembered as a project belonging to Peto and Grissell, who built the later stages). Aptly enough, Henry's son William later frequented the Houses of Parliament in a different capacity – as an MP in the 1850s.6

Not everyone made a fortune. Sadly, even Thomas Piper ended up needing the benevolence of the Society. After his business was bankrupted in 1862, his peers in the Society sympathetically voted to donate him an extremely generous £500. They also appointed him as the Society's first paid Secretary, an office he held until his death in 1873.7

Building bridges

However, most of the Society's charitable efforts were designed to help the workers and their families. Despite their sometimes pugilistic attitude towards labour relations, London's leading builders took a rather more benign view of their workers' welfare when union matters were not involved.

Many leading members of the Builders' Society were staunch supporters of the Builders Benevolent Institution, which was founded in 1847 as a separate concern (It still exists today and has a close relationship with the work of the Federation of Master

Builders.). However, the support efforts were not totally without self-interest, as a subsequent report of 1848 in the 'Builder' makes clear:

"We are glad to perceive that the list of stewards for the anniversary dinner... is still on the increase... our desires, however, on this point are not very easy to satisfy, and we therefore still exhort every one of our influential metropolitan builders who has not yet found time to add his own name to the list, forthwith to do so. They do not overlook the fact, we hope... that this is peculiarly the workman's cause – a cause which at this moment demands that every iron shall be thrust in the fire, to solder and amend those evils which otherwise threaten the entire disintegration of the social relationship between master and workman. And that a builders' charitable institution is, above all, important in the present crisis, we have only to look to the deplorable state of our continental neighbours to be too well assured..."8

As this excerpt suggests, these were turbulent times in Europe. 1848 was known as the 'year of revolution' and also saw the publication of the Communist Manifesto. For the master builders (and any others who saw themselves among the elite), it was important to maintain the status quo in Britain; complacency couldn't be allowed to dilute efforts to keep the working classes content with their position in society.

The Builders' Society also supported the establishment of a provident and friendly society for building and engineering workmen.9 A meeting between leading architects and builders was held at the London Guildhall on the 11th December 1849, with a view to making provision for their workers' old age. The builders in attendance were Thomas and William Cubitt, Alderman Lawrence, Henry Lee, Thomas Piper and a Mr Baker.

By the 1850s, however, industrial unrest was growing once more. Campaigning for the Nine Hours Movement, workers wanted a shorter working day for the same amount of pay. Initial discussions took place directly between the Society and the men, but no agreement could be reached. Then the Society announced that "a meeting of the Metropolitan Builders would be convened by the masters to consider the subject."10

On the 20th April, the Metropolitan Builders passed a resolution, proposed by Charles Lucas and seconded by Myers, not to agree to the nine hour day. Mr Dummage (of Cubitts) proposed, and Kelk seconded, a resolution that such acquiescence would amount to a

5   Satoh, A. (trans. Morton. R.) 1995. *Building in Britain: the origins of a modern industry.* Scolar Press: Aldershot pp. 68–9

6   Ibid. pp. 61–2

7   Houldsworth, H. K. 1982. *The Builders' Society 1834–1884.* Unpublished dissertation: Trent Polytechnic, Nottingham p. 30

8   *The Builder 1848* Vol. 6 Jul 8 p. 334

9   Anon. 1850. *Proposed Builders and Engineering Workmen's Provident and Friendly Society.* In The Labourers' Friend, January pp. 6–8

10 Shaw Lefevre, G and Bennet, T. R. *Account of the Strike and Lock-Out in the Building Trades of London in 1859–60* In Report of the Committee on Trade Societies, appointed by the National Association for the Promotion of Social Science, presented at the Fourth Annual Meeting of the Association at Glasgow, September 1860 London: John W Parker and Son pp. 52–77

tax on the public – presumably because of the number of public commissions – and set a bad precedent for other trades. This was also passed.

The operatives wrote one last letter, asking the Society to reconsider. George Wales answered, reminding the workers that "the Society felt that the question was too important to be dealt with by them, and it was therefore determined to convene a meeting of the Builders of the Metropolis… it appears to this Society, that the concluding paragraph of those resolutions does very distinctly answer the request contained in your letter."[11]

It is interesting that Wales says it was 'too important' for just the Society's consideration. This suggests that the creation of larger employers' organisations related to the Society indicated a desire for some form of democracy for the most important decisions – without it diluting their otherwise exclusive club.

By July, the men working for Trollope had gone on strike, "irritated" when he dismissed a worker for presenting their demands. In retaliation, the masters started a lockout in August while remaining committed to seeking a way of letting non-union men back in. This led to the resurrection of the hated 'document', signing away union rights in return for work. As before, the masters offered to assist with any "legitimate Benefit Society", but the newly-formed Central Association of Master Builders (CAMB) was resolved not to employ unionists. Masters such as Trollope, Cubitt and Piper reopened their doors while insisting that signing the "declaration" was a condition of re-entry. Only a small proportion of workers returned.

Once again, the 'document' proved a failure. Men got jobs with smaller concerns, or travelled out of London to work. Others were supported by the solidarity of their fellow unionists.[12] Trades from the hatters to the farriers to the pianoforte-makers contributed funds. Nevertheless, the men and their families suffered for their principles.

However, not all disputes were as widespread, bitter or polarised. George Howell, a bricklayer who went on to become a Liberal MP and a successful author[13], recounted an incident when he was working as foreman for Dove Brothers in the early 1860s. The labourers suddenly made a completely unreasonable and unexpected demand for a rise – one morning, at seven o'clock – and threatened to down tools.

Howell told Dove, who told him to sack the ringleaders and pay the rise, but only "to those who were worth it". As Price notes, this is an example of "a group of labourers acting in a tactically

suitable way – the early morning was an appropriate time to seek an increase because if labour conditions were right and the work important the employer would be obliged to grant it – and Howell, the union man, only two years off the blacklist for his role in the 1859 and 1861 strikes, enforcing the discipline of the employers."[14]

This anecdote is a useful reminder that not all disputes were a bi-partisan stand-off. A range of interests, not all of them mutually exclusive, were part of the industrial relations of the time, just as they are today.

An extension of interests

From the 1840s, the Builders' Society was also active in providing commentary on draft legislation. Over the coming decades, the Society passed judgement on everything from Conciliation Boards to Traffic Management, as well as mounting vigorous opposition to the Metropolitan Buildings and Management Bill and the Employer Liability Act in the 1870s.[15] By 1871, the Society had also collaborated with the Royal Institute of British Architects (RIBA) to create a new form of contract, which the Society was bound to use.[16]

However, an invitation from RIBA to revisit this work in 1878 was first of all rejected and then subsequently accepted[17] following transient concern that the National Association of Master Builders (later the National Federation of Building Trades Employers – NFTBE) was already looking at these issues. The episode demonstrated a lack of clarity over roles and relationships, which would become increasingly problematic in the decades to come.

In 1879, Stanley Bird (a member of both CAMB and the Builders' Society) suggested that CAMB sought incorporation. When this proved unsuccessful, Bird promoted the idea of the Builders' Society achieving incorporation instead.[18] This did work and he became the first president of the new Institute in 1884. Although little about Stanley Bird's professional career seems to be readily available, we do know that he was a successful London builder and also a vice-president of the National Rifle Association.

Another early president, George Burt, left a far more prominent legacy. As the nephew and successor of John Mowlem, he built on the success of his uncle's business by undertaking contracts such as Old Billingsgate Market, the Victorian Embankment and the City of London School. He and his colleagues in the new Institute of Builders (IOB) were immensely active and influential in the trade at that time. For example, the IOB secretary, Thomas

11 Ibid.
12 Anon. 1860. *Builders' Combinations in London and Paris* in The National Review XXII, October p. 321
13 Postgate, R. 1923. *The Builders History London: The National Federation of Building Trade Operatives*. p. 221

14 Price, R. 1980. *Masters, unions and men: Work control in building and the rise of labour 1830–1914*. Cambridge University Press.
15 Houldsworth, H. K. 1982. *The Builders' Society 1834–1884*.
16 Ibid. p. 27
17 Ibid. p. 28
18 Powell, C. 1997. *Diligently and Faithfully*. Ascot: Chartered Institute of Building

Costigan, also acted for the London Master Builders' Association (LMBA) and the Builders Benevolent Institution, as well as all the London Area Consolidation Boards.

The NFBTE itself could list Bird, Colls, Sapcote and Rider among its past presidents, and had many members and interests in common with the IOB (some of which dated back to the Federation's previous incarnation as the National Association of Master Builders). Other areas where interests overlapped included the introduction of the Employers' Liability Bill in 1881, and the addition of industry representation on key boards of enquiry.

"Perhaps as a consequence," says PJ Spencer, a future Institute secretary, "the first leg of the Institute's primary purpose went: a leading member of the Institute, upon appointment to Federation Office, took the Institute's contract negotiations into his own hands and began the process of handing the business of the Institute over to the National Federation to conduct and compromise."[19]

Despite this loss of control in contractual matters, which has persisted ever since, the Institute was making significant strides in another equally important area: educational affairs.

The Institute had set up a committee for education in 1905, and subsequently become involved in technical education for tradesmen and apprentices. Labour shortages occasioned by the First World War gave the IOB even greater incentive to promote an apprenticeship scheme, which was approved by the Board of Trade.

Post-war reconstruction

The Great War ended in 1918, the same year that the Institute moved to 48 Bedford Square, London – which remained as its headquarters for the following 50 years. Membership began to climb sharply too, more than doubling from 251 in 1921 to 561 in 1924. The 1920s also saw the foundation of the Architecture and Surveying Institute – now the Faculty of Architecture and Surveying – within the CIOB, as well as the start (in 1923) of examinations for entry to the Institute – although exceptions would continue to be made for any suitable candidates considered "above examinable age". The Institute applied for its Royal Charter in 1924, but failed due to its requirement (established five years earlier) for members to also join the NFBTE. The rule was subsequently removed. 1924 also saw the appointment of Percy Spencer, the first full-time secretary since Thomas Piper.

Spencer refused to accept incursions into the work of the Institute without mounting a ferocious defence. For example, during World War Two, the NFBTE and the National Federation of Building Trade Operatives (NFBTO) had become the bodies the government

consulted for emergency policy-making. In a confidential report of 1946, Spencer complained of "two serious defections" from members of the Institute's Council to the National Federation – their crimes being the promotion of the trade association's views on matters he saw as within the Institute's remit.

Professionalism, as opposed to commerce, was Spencer's clear aim. He felt that the domination of trade interests had hindered the development of the Institute since its inception. He even alleged that, since the incorporation had been prompted by the failure of CAMB to achieve the same objective, the IOB's "constitution was drafted for it rather than by it… for forty years – apart from sporadic interest in the crafts expressed in medals and prizes – it devoted itself solely to employer interests."[20]

While Spencer's insights are not without foundation, they appear to be somewhat overstated. For instance, the Institute's "sporadic interest in the crafts" had been far more than mere "medals and prizes" – it had devoted a large amount of time and effort in the early part of the century to improving trades education. Similarly, the relationship with the NFBTE had been clearly unhelpful in some respects, but was absolutely necessary in others. By trying to demonstrate the Institute's strength and independence, Spencer was helping to erode vital partnerships and leaving the Institute in danger of becoming virtually moribund.

A new dawn

By the mid-fifties, Spencer was under mounting pressure to retire. The 'Builder' welcomed his resignation with the comment that it "marks the end of a chapter in the Institute of building education and the beginning, it is hoped, of another. . . Failure by the Institute to come to terms with other bodies closely connected with and interested in building education has caused vexation and frustration."[21]

Things could only improve and, with the young and enthusiastic Dennis Neale at the helm as the Institute's new Chief Executive, they did. By 1959, Mr G O Swayne FIOB OBE was able to give a presidential address charting the step-change in progress that had taken place over the previous few years:

"Some ten years ago the Institute was under heavy fire of criticism from its own members, from the industry at large, from educationalists and from the technical press, for its general inertia... It was urged to put its house in order, to broaden its membership conditions, to end "special case" entry, to modernise its examination syllabuses with a greater accent on management, to publicise its purposes and intentions, to decentralise its activities and to inaugurate a Board of Building Education. Gentlemen, in five years it has accomplished most, if not all of these things."

19 Spencer, P. J. 1946. *Confidential Secretary's note upon the Institute, for consideration at a meeting of the Administrative Committee on Wednesday 28th August 1946.*

20 Spencer, P. J. 1946. Ibid.

21 Quoted in Powell, C. 1997. *Diligently and Faithfully* p. 34

Professor John Bale PPCIOB remembers that the new curriculum introduced at that time, and accompanying texts such as the Building Management Notebook, were "really seminal work. It amounted to the first real attempt at defining the discipline of construction management."

The members who were driving this enormous achievement (including Sir Peter Shepherd and Professor John Andrews) would continue to have an enormous influence on the development of the Institute over the following decades. Their impact was about to reach its height with the inauguration of Peter Shepherd as President, from which position he helped revolutionise the Institute in a number of very significant ways, not least in its change of name.

"The name 'Builders' made it sound like an association purely for proprietors and directors," says Dennis Neale. "That was a bloody nuisance. We couldn't progress. But 'Building' was all embracing – by changing the name, we recognised sub-specialisms and the growing educational sector."

A change of a different kind was necessitated at the end of the sixties. The Institute's longstanding home in Bedford Square (leased throughout the decade at the accommodating rate of six shillings and ha'pence per square foot) was under new ownership and due for a lease renewal. The owners offered terms of five pounds per square foot, which quickly focused attention on a serious search for different premises. The Institute's new home in Ascot took it out of central London, which had drawbacks in terms of ease of access but also advantages in terms of offering much-needed room for expansion.

In 1970, another milestone was reached when the Institute was granted charitable status. As the seventies drew to a close, Dennis Neale was drafting the application for incorporation by royal charter. By 1980, the Chartered Institute of Building, and its Chartered Builder members, had finally gained the societal recognition of royal approval and consequently experienced significant growth.

International expansion

The Institute has long had members outside of the UK, but its emergence in the late 1990s as a truly international organisation can be ascribed to two factors.

Firstly, the internet revolution and the advent of e-mail made quick and inexpensive global communications a reality. Michael Brown, Deputy Chief Executive, believes that these changes "opened things up dramatically. It allowed us to build relationships that just weren't possible ten years earlier."

Secondly, Paul Shepherd PPCIOB (son of Sir Peter) became Chairman of CIOB International and proposed a system whereby subscription payments from overseas were ring-fenced for international development activity. This innovation served a double purpose: it helped overcome some members' perception that international travel by senior members and staff was

both expensive and undesirable, and it also allowed a planned programme of development and growth – including the opening of a string of international offices.

In the early days, however, no one could have foreseen the future success of the Institute's international expansion or the level of influence to be achieved. One contributory factor has been the changing socio-economic and political circumstances in China. These developments have encouraged the Chinese government and the country's construction industry to build a strong relationship with the CIOB, since the subject of construction management needed a complete rethink of the curriculum.

"It's hard to say anything which effectively communicates the sheer scale of the Chinese industry," says Michael Brown. "There are 40 million people employed in construction in China, and the CIOB has provided real leadership to the entire industry via a small group of very senior people who have become members. The Department for Business, Enterprise & Regulatory Reform could not believe how much we had achieved without the direct support of UK government."

Looking to the future

A new Chief Executive, Chris Blythe, took the helm in January 2000 and the start of a new century marked the beginning of a rejuvenated Institute. He has swept away the last vestiges of the "Old Boys' Club" that sometimes characterised the body in the past, and set in motion substantial changes to its governance too – making it more suited to agile decision-making.

The profession itself has risen to a new prominence within the built environment disciplines. Undoubtedly, contractors have developed a leadership position – in small part due to changing working and contractual arrangements – that wouldn't have been anticipated by the majority of the architectural community 40 years ago. Indeed, Chartered Builders all over the world are leading a myriad of important and useful construction projects.

As Roger Flanagan PPCIOB commented in his introduction to the CIOB's 2007 Construction Leaders publication, "For every exceptional personality featured here, there are thousands of gifted people working to deliver a better built environment… This tribute to some of our leaders, therefore, is also a testament to the contribution of every man and woman who brings professionalism and commitment to their career in construction."[22]

No better words can be found to introduce the following profiles, which provide at least a flavour of the extraordinary achievements of this most modest group of professionals: The Chartered Builders.

22 Flanagan, R. in Collier, C. 2007. *CIOB Construction Leaders* Chartered Institute of Building: Ascot

Thomas Cubitt {1788–1855}
William Cubitt {1791–1863}

Founder Members of the Builders' Society

Born in the Norfolk village of Buxton, Thomas and William Cubitt were descended from an obscure line of farmers. However, around 1792, their father Jonathan moved the family to London, where their two younger brothers, George and Lewis, were born.

When Jonathan died in {1806}, eighteen-year-old Thomas was left with three younger brothers, an elder sister and his widowed mother to support. Thomas was at that time a journeyman carpenter, and it would be said later that "The uncertainty of such a position made a deep and lasting impression on his mind, and stimulated him to unceasing exertion, in order to obtain a more independent position."[1]

To set himself on the road to this independence, he became ship's carpenter on a voyage to India. Using the money he earned to start his own carpentry business when he returned to London, he established himself in his first premises in Holborn. Not long after the owners of the Russell Institution, who had recently purchased the building for use as a subscription library, found that they had serious problems with their roof. This was unexpected in a relatively new building, but further investigation found that it was "not only in bad condition, but formed of improper material"[2]. A complete replacement was required. Disinclined to use the workmen responsible for the previous construction, they hired the young Thomas Cubitt. He completed it satisfactorily and rapidly, according to his biographer, Hermione Hobhouse[3]. The glowing reports he won for this job appear to have been crucial to his winning another major project, the construction of the London Institution.

By the time he was awarded this contract, Cubitt had already succeeded at a small, speculative venture in Chiswick and was styling himself "Builder" rather than "Carpenter". Significantly, this implies that he already had experience of subcontracting work to other trades.[4]

These early years of trading saw William Cubitt return from a spell in the navy to join his brother's business. However, the untimely death of the third brother, George, at around the same time may have contributed to the decision that Lewis, the youngest, would become an architect's apprentice rather than join the family firm.

Thomas's new clients, the London Institution, had been in temporary premises in Cheapside. After much deliberation, they decided to build their own impressive headquarters where "the order is Corinthian, and a modification of the celebrated example taken from the Temple of Vesta at Tivoli."[5]

The budget for this ambitious project was £20,625 and there was a stiff penalty clause for late completion. Cubitt concluded that subcontractors were too unreliable and that he needed his own men, even if that obliged him to pay more. That's why this project is held to mark the advent of the major contractor.

It is difficult to convey how revolutionary this approach seemed at the time. According to an obituary, "With a promptness and daring which marked Mr Cubitt's character in all future works, he immediately purchased a piece of ground on the east side of Gray's Inn Road, commenced a series of workshops… and engaged gangs of carpenters, smiths, plumbers, glaziers, painters, bricklayers, etc with a foreman to each class. This bold and hazardous plan was a novelty in London… and provoked much speculation, with some envy…"[6]

However, he still had to rely on an external architect, who failed to spot problems with the foundations. Coupled with late changes in the brief, this meant that the cost spiralled. Some of this was borne by Thomas, who concluded that "the best mode of proceeding with new buildings was to be independent of architects."[7]

Thomas and William both became involved enough in design to be described as 'architects' in the press, but they disliked the title and neither ever adopted it. Rather, as William became prominent in civic society, he would sign himself 'William Cubitt, Gentleman'. Thomas, in contrast, always styled himself 'Builder'.

This was not the only difference between the two. William was more cautious, and managed the contract side of the business. Thomas, however, was more comfortable with the risky, speculative life of a developer. The London Institution contract had introduced Thomas to wealthy backers, who provided him with initial capital for the immense speculative building projects that followed.

1   Britton, J. 1856. *Obituary, Thomas Cubitt. The Gentleman's Magazine.* February. London: John Bowyer Nicholls and Sons p. 202
2   Ibid.
3   Hobhouse, H. 1995 2nd Ed. *Thomas Cubitt, Master Builder.* p. 5
4   Ibid. p. 6

5   Upcott, W. R. Thomson, and E. W. Brayley. 1835. *A Catalogue of the Library of the London Institution: systematically classed.* Preceded by an Historical and Bibliographical Account of the Establishment. p. 15
6   Britton, J. 1856. *Obituary, Thomas Cubitt.* p. 203
7   Hobhouse, H. 1995. *Thomas Cubitt, master builder.* p. 17

{1806}

The Arc de Triomphe
is commissioned by
Napoleon Bonaparte.

{1823}

Work begins on the
British Museum in
London, designed by
Sir Robert Smirke.

{1823}

William Strickland built
Saint Stephen's Church in
Philadelphia, Pennsylvania,
one of the first Gothic
revival buildings.

{1827}

Work begins on the
Athenaeum Club in
London, designed by
Decimus Burton.

By the 1820s, the brothers were working on developments in Bloomsbury that would transform the Duke of Bedford's fields to the south of Russell Square. Thomas would take long leases, usually for 99 years (the entails – or 'fee tails' – placed on the lands of the aristocracy often prevented outright sales). This gave him the land at a peppercorn rent for the first few years, before the fee climbed sharply as the lease progressed. Hobhouse, who believes that their younger brother Lewis was involved with their designs at this time, finds these early streets "among the most elegant and artistically satisfying of the Cubitt developments"[8].

Whatever the importance of Lewis's possible input into the aesthetics of the development, the plans submitted by Thomas in {1823} show that he was already committed to an approach to building that would be the hallmark of his career.

His accompanying letter explained: "You will observe the widths of the Streets and of the Square. I have made them wider than was shown on your plan but I consider it will not interfere with any former arrangements in a way to prejudice the Estate, but on the contrary will make it appear more healthy and will be filled with the best class of occupiers and I think there can be no doubt but that it will prove most to the advantage of his Grace by giving a superior character to his Estate by making the Communication with the New Road better than the adjoining neighbourhood."[9]

Thomas was the antithesis of the jerry-builder. His instincts were always to provide the highest standards of workmanship, good roads, good drainage and generous amounts of space. In the early years of the Bloomsbury development, this approach paid off handsomely. The leaseholds on his houses sold well, and business was good. Towards the end of the 1820s, however, a combination of harsher economic conditions and the beginning of fashionable society's migration from North London to the West End were causing problems.

He asked the Estate office if he could turn three unsold houses into a seminary for young men. The response from Adam, the Estate auditor, was quite caustically unenthusiastic:

"I don't understand what he means by young men 'who having been educated at Hertford', come 'to finish their education at the University of London'… I am not aware of any young men being educated at Hertford… I cannot therefore understand what class of young men Mr Cubitt proposes to have in the house."[10]

Adam's letter continues by asserting that Thomas's idea is a risky suggestion, which could be very bad for the neighbourhood, and that "nothing but a strong desire to help Mr Cubitt's large speculation would make it to be entertained."[11]

The strong desire to help was evidently present, because the seminary went ahead. Various shrewd deals were done with the Estate to mitigate the problems created by the changed market conditions, and Thomas survived unscathed. By this time, Thomas was also creating even more ambitious developments in Belgravia, so he would benefit from London society's move westwards even as he lamented the difficulty in selling the Bloomsbury properties.

Nevertheless, the risks Thomas undertook were huge, and this may be why the brothers divided their operations in two in {1827}. Hobhouse[12] finds their decision puzzling from a business perspective, insofar as the risk from the profitable speculations had been mitigated by the contract building projects. The split exposed Thomas to greater risk and dependence on bankers, whilst depriving William of potential profits. The biographer therefore suggests that Thomas's autocratic nature may have contributed to the division.

In the days before the 1855 Limited Liability Act, an alternative possibility is that the brothers believed that separating the safer contracting business was the best option. Certainly, William helped his brother financially in 1833, which suggests that the contracting side could still informally underwrite the speculation should the need arise. What's more, had the worst happened in Thomas's developments, William's contracting business would have survived to benefit them both.

In an age where family networks were the greatest bulwark against hard times, it was clearly better to ensure that at least one portion of the business was insulated against risk. There appears to be no evidence that the break was in any way acrimonious, and they

---

8 Hobhouse, H. 1995. *Thomas Cubitt, master builder.* p. 66
9 Ibid. p. 67

10 Ibid. p. 76
11 Ibid.
12 Ibid. p. 96

Buckingham Palace | London, United Kingdom

{1831}

The Dugald Stewart
Monument in Edinburgh,
Scotland is completed.

{1831}

The Bridge of Sighs,
Cambridge, England
is completed.

{1844}

The Scott Monument in
Edinburgh, Scotland is
completed.

{1855}

Palais d'Industrie was
built for the Exposition
Universelle World Fair
in Paris.

made informal and logical decisions when splitting the assets
without recourse to law. This suggests that the decision was mutual,
and it seems entirely possible that this is just another example of
the brothers' shrewdness.

The younger brother Lewis then went into partnership with Thomas.
But by {1831}, William's contracting business was expanding
and Lewis returned to his company at Gray's Inn Road to assist.
While never quite in the same league as his brothers in terms of
achievements, Lewis was a well-regarded designer. He later set up in
private practice, and developed a small but successful speculation of
his own before later becoming a railway architect. He is remembered
as the architect of Kings Cross Station, and finished his career
working for the Great Northern Railway.

In 1831, however, William and Lewis were busy building somewhere
to house a different kind of mechanical progress – Mr Babbage's
Calculating Engine. The project was perceived as urgent, according
to a report sent by the architect, Decimus Burton, to the project's
paymasters at the government Works Department. Burton claims
that "in consequence of the urgent representations which were
made by Mr. Babbage &c. of the importance of having the works
completed with all possible dispatch, I therefore directed Messrs
Cubitt to use every exertion in forwarding the Buildings…".13

It appears that the client was being gently prepared for the costs
of accelerated working, but the haste was ultimately unnecessary.
Mr Babbage's engine was not successfully realised until 1991, when
an experiment by the London Science Museum proved that it
could not only be built, but could also perform calculations to 31
decimal places. The reasons for it never being built successfully in
the inventor's lifetime remain subject to conjecture. Nevertheless,
the brothers were working on what was almost a project of great
historic importance: the building that should have housed a
Victorian computer.

Around this time, they also built Covent Garden market and won
contracts for a great number of very substantial townhouses. Thomas,
meanwhile, was pressing ahead with his enormous developments,
which quite literally shaped the London we know today.

As Hobhouse points out14, before the Metropolitan Buildings
Act of {1844}, builders were at liberty to set their own standards.
Thomas chose to build houses of substance, but this duty was not
imposed on him by law. Insofar as there was regulation, it was
imposed by the offices of the great estates, keen to safeguard the
quality of their holdings. But the existing modus operandi was no

longer sufficient to deal with a metropolis exploding in size, and
this led to a series of commissions for enquiry and improvement.
As Cubitt progressed his great developments in Belgravia, Pimlico
and beyond, he also regularly gave evidence to those seeking to
improve the quality of London's 'built environment'.

The term sounds almost anachronistic when discussing the
Victorian age, but it was a truly environmental problem. The
smoke and smells of bad chimneys and bad drains choked the city,
and the Victorians' way of conceptualising the issue bears striking
similarities to our own discourses about such problems.

In 1842, Chadwick published his report on The Sanitary Condition
of the Labouring Population of Great Britain. Thomas's evidence
makes it clear that, despite his great entrepreneurial spirit, he was
in favour of government intervention at least to some degree. He
thought that public officers should enforce standards relating to
sewers and drains, and that smoke was the cause of much misery
to Londoners. He had no illusions about the attitude of the culprits
or the likelihood of self-regulation:

"A man puts up a steam-engine, and sends out an immense
quantity of smoke… where his returns are £1000 a month, if he
would spend £5 a month more he would make that completely
harmless, but he says 'I am not bound to do that', and therefore
he works as cheaply as he can, and the public suffer to an extent
beyond all calculation."15

It is interesting to note that Chadwick's report describes the
situation outlined by Thomas as an "oppressive tax occasioned
by… carelessness". This is a term that was also used to describe
the increased wear and tear on linens in urban areas, which were
quickly filthy and thus more frequently washed. Although they
had not moved to the idea of taxing the polluter, the concepts of
tax and environmental problems were already being linked in the
discourse of the 1840s. It suggests a holistic understanding that
there was a very real cost to environmental degradation, and that it
was having an impact on quality of life.

In his yards, Thomas used furnaces that consumed their own
smoke while the smoke from William's premises in Gray's Inn
Road went through a condenser and into the sewers. Thomas
argued strongly that the extra cost of manufacturing with clean
engines would pay dividends through the increased lifespan of all
the household items being speedily ruined at that time, using what
Hobhouse terms 'an early piece of cost benefit analysis'.16

---

13 Letter from Decimus Burton to Sir B.C. Stephenson, Surveyor General,
6th February 1832
14 Hobhouse, H. 1995. *Thomas Cubitt, master builder.* pp. 103–114

15 Chadwick, E. 1842. *Report to Her Majesty's Principal Secretary of State
for the Home Department, from the Poor Law Commissioners: On an
Inquiry Into the Sanitary Condition of the Labouring Population of
Great Britain: with Appendices.* London: HMSO
16 Hobhouse, H. 1995. 2nd Ed *Thomas Cubitt, master builder.* pp. 434–5

{1855}

The Old Stone Church was
built in Cleveland, Ohio.

Thomas's interest in the general quality of life in London would also make him an enthusiastic advocate of public parks for the poor. Eventually, this landed him the honorary role of an examiner for District Surveyors, a new position created under the Metropolitan Buildings Act of 1844.

At around the same time, he received an appointment with far more social cachet – Queen Victoria and Prince Albert chose him to remodel their newly purchased country retreat, Osborne House, on the Isle of Wight. According to Hobhouse[17], this caused some comment since Thomas was not an architect. Nevertheless, he was one of the few builders confident enough to deliver a project with a fixed cost and schedule. He became well-liked and trusted by the Royal Family, who also used him on the project to remodel Buckingham Palace. In another incredibly modern innovation, Thomas offered to run the contract with an open account book, undertaking to draw no more than seven per cent profit on the project.

When Thomas died in {1855}, Queen Victoria wrote in her diary that she was much grieved by the loss of an "excellent and worthy man"[18]. Many of her more humble subjects were equally saddened by his loss.

In the words of his obituary, "Through life he has been the real friend of the working man; and among his own people he did much to promote their social, intellectual, and moral progress. He established a workman's library; school-room for workmen's children; and by an arrangement to supply generally to his workmen soup and cocoa at the smallest rate at which these could be produced – assisted in establishing a habit of temperance, and superseding, to a great extent, the dram-drinking which previously existed among them."[19]

He was fondly remembered for his reaction to a major fire at his yard. Despite his own huge losses, his first thought was to reassure his workmen that they would be back in business soon, and to subscribe £600 towards new tools for them.[20]

Perhaps his most important legacy, however, is his contribution to civic development. He was an early proponent of comprehensive planning for good sewerage, provided crucial support to the project to create the Chelsea Embankment, argued against polluters and left a legacy of a well-proportioned and generously laid out built environment for large swathes of London. Much of it still stands today. A developer by inclination, he was also a contractor to the greatest in the land.

Although drawn to contracting, William was immersed instead in a major development project by the early 1850s. Although he built more modest houses, on a more modest scale than Thomas, the portion of the Isle of Dogs known as Cubitt Town is named after William, not his illustrious brother. William's involvement in the project was timely – the value of this low-lying land was diminishing, and investment was required for draining and embanking. According to the Survey of London, "The area had no road access and it was thought at that time that the nature of the ground made it unlikely that a railway could ever be built across it. Moreover, the foreshore was being steadily eroded by the wash caused by steamships, traffic which was obviously going to increase."[21]

William's participation, then, was clearly beneficial to the estate-owners. As with Thomas's speculations, he negotiated a peppercorn rent in the first instance, which rose as the land was developed.

William also showed far more interest in formal involvement in civic society than his more independently-minded brother. He was an Alderman and a magistrate, served as deputy-lieutenant of Hampshire, and was the MP for Andover (1847–1861). He resigned that post to become Lord Mayor of London, a position he would retain until shortly before his death.

17 Ibid. pp. 377–8
18 Ibid. p. 473
19 Britton, J. 1856. *Obituary, Thomas Cubitt.* p. 202
20 Walford, E (Ed.). 1856. *Harwicke's Annual Biography.* London: Robert Hardwicke p. 297

21 Hobhouse, H. (Ed.) *1994. Survey of London: volumes 43 and 44:* p. 495

Osborne House | Isle of Wight, United Kingdom

His political standpoint seems uncomplicated, and he was probably more interested in practical concerns than in lawmaking. He was Conservative, but favoured free trade, according to his obituary. The one political anecdote which is recounted there suggests that the intricacies of politics were never his over-riding concern:

"On one occasion he voted for the abolition of Church rates, but it was said that by accident he went into the wrong lobby, and did not discover his mistake in time: this appears probable from the fact he never repeated the vote."[22]

Comparing the obituaries of the two brothers, one of the most striking things is how differently they must have chosen to present themselves. Thomas was very much the self-made man, always described as the son of a labourer or similar. He never ceased to identify with the working people ("a source of pride rather than humiliation with him", according to a contemporary[23]) and turned down all titles and honours he was offered. His son George became the first Baron Ashcombe, a title similar to the one he might otherwise have inherited, but Thomas remained plain Mr Cubitt all his life.

William's obituary, by contrast, describes him as the son of Jonathan Cubitt, Esq. It seems likely that, as the man who would be Lord Mayor of London in the year of the second Great Exhibition (and thus responsible for welcoming senior foreign dignitaries to one of the world's greatest cities), he preferred to emphasise his respectability over his entrepreneurship. In any case, it is probable that both narratives are true – Jonathan would have had some claim to the title of a gentleman, as there are records of him voting, but his legacy must have been negligible.

It is also noticeable that the posthumous assessments of Thomas were concerned entirely with his developments (and benevolence to his workforce). In contrast, William's building career is hardly mentioned after his death, except in tributes from the architectural profession. To the general reader, it was far more important that he was Mayor. He was also President of St Bartholomew's hospital and Prime Warden of the Fishmonger's Company, an honour perhaps not unrelated to the fact that he had built their Hall in the early 1830s. However, one of William's greatest achievements was completely separate from his activities within the building trade – activities that triggered an outpouring of popular grief at his death.

This widespread outbreak of deeply felt mourning did not take place, as might have been expected, in London, but miles away in Lancashire. Following a suggestion in the local press, a muffled peel of commemoration was rung in Manchester, Wigan, Bury, Blackburn and a multitude of small towns across the north-west.

At Ashton-under-Lyme, it was reported, "the community turned out in mourning as if for the funeral itself". This was not an official civic procession, but a demonstration of grief from the people, and it left an impression on contemporaries. The poor gathered in large numbers, "many of them carrying lighted torches, which gave a very novel and funereal effect to their ranks as they moved through the drizzling wet of a dark and boisterous evening… a more beautiful and impressive scene could not have been arranged on so short a notice by any court master of ceremonies… Many who witnessed the deep feeling of the assemblage will not soon forget it."[24]

So, how did a London builder become the object of such affection to the Lancashire poor? Not only physically distant, but a world away in culture, they had likely never seen a Cubitt building.

The answer lies in William's work as Lord Mayor of London, when he came to their aid in the Lancashire Cotton Famine. The incident is worth recounting in some detail, because it tells much about his character and way of working.

The great mills of the north-west of England depended on cotton from the American South. When the American civil war broke out, the southern ports were blockaded by the Union. At first, it seemed that stocks would last, but it quickly became clear that there would be no swift resolution to the conflict. Unemployment and privation became widespread in these English mill towns.

The matter was brought to the attention of William and, in his capacity as Lord Mayor, he formed a Mansion House Committee. William already had experience of fund-raising in the wake of the Indian Mutiny and when initiating subscriptions for the Albert Memorial, a project he led in its early stages. It was reported that more than half a million pounds had been raised for the Lancashire workers by the time of his death – a monumental sum of money which provided help to thousands.

Through the efforts of his London committee, money reached the workers at great speed, outpacing local relief efforts. Despite this, his work for relief aroused great controversy. The Manchester

22 1864. *Obituary, William Cubitt. The Gentleman's Magazine.* January. p. 121

23 J.W.B. 1856. *Clapham Park and Mr Cubitt. The Gentleman's Magazine.* April. p. 382

24 Jones, W. H. 1863. *The Muffled Peal: A Sermon Preached in the Parish Church of Mottram in Longendale, on Sunday afternoon, November 8 1863.* Being the Sunday after the funeral of the late William Cubitt, Esq MP. London: Hatchard and Co. pp. 23–4

Committee wanted to distribute the monies from London, and felt that the generosity from the South was sometimes misplaced. In an age where "outdoor relief" (that outside of the workhouse) was not the norm, their approach to giving was at odds with the local relief committees. While the local bodies were more anxious to establish the credentials of the needy, London gave relatively freely. Henry Jones, a vicar who preached a memorial sermon for William, said that the local charities learnt from London's example "to treat their suffering neighbours… as honest men until they found them to be rogues."[25]

Over-zealousness at local level nearly led to serious civil disturbance at Stalybridge, when a dole cut coincided with the introduction of a ticket system instead of money. Riots ensued, and following a deputation from the local clergy, London voted them an extra £500 to defuse the situation. At a time when the French Revolution was still just within living memory, and voting rights for working men were still two decades away, this was not greeted with universal approval. Rioters were not to be placated.

The Central Manchester Committee in any case supported the cut in dole, and deemed that "The interference of the Mansion House Committee was dangerous, although no harm came of it."[26] However, more liberal commentators in the clergy thought the actions of the Committee prevented a tragic and sustained civil disturbance, where at worst British troops could once again have been ordered to fire on their countrymen. The Cotton Famine riots could have become a latter-day Peterloo.

In the words of one grateful mourner: "What has been, and what might have been had the Lord Mayor and Mansion House Committee yielded to the influential overtures made them to give up their fund to the Manchester Central Committee, makes one feel truly grateful that they stood firm, unbending, and determined to do their duty."[27]

The people of Lancashire knew him only as Lord Mayor, not as the successful builder. But to those who understand his career, there can be little doubt that his pragmatism, organisational skills and self-belief underpinned both his building and his civic contribution.

The abilities that made it possible for him to feed the hungry so effectively were, in essence, the exceptional project management skills of a trail-blazing major contractor.

25 Ibid. p. 17
26 Arnold, R.A. 1864. *The History of the Cotton Famine: From the Fall of Sumter to the Passing of the Public Works Act.* London: Saunders, Otley, and Co.
27 Anon. Quoted in: Jones, W. H. 1863. *The Muffled Peal.* London: Hatchard and Co. pp. 23–4

Sir Samuel Morton-Peto
{1809–1889}
Founder Member of the Builders' society

Although it is always important to place historical figures within the context of their time, perhaps the most noticeable thing about this eminent Victorian is the extent to which his views accord with those of today.

While the terminology has changed – today we talk about 'transparency', whereas Peto was 'anti-bribery' – his attitudes were strikingly modern in many respects. A man who combined business with a long career in public life as a Member of Parliament, he consistently sought to raise standards of probity, welfare and business efficiency. As with so many great Victorian builders, his achievements were staggering and executed with a confidence and energy that reflected and defined the entrepreneurialism of the age.

Born in 1809, Peto showed a talent for drawing and handwriting during his early schooling – the latter skill making him much in demand among the school's servants for his services as a letter writer. He retained an interest in education throughout his life, regretting his limited linguistic skills as a businessman with global interests. In his sixtieth year, during a stay on the continent, he wrote of "the absolute necessity of our boys being thoroughly up in French and German. I find the boys here and in Vienna far better linguists and better mathematicians, and better acquainted with scientific matters generally, than our English boys, and the area is really so much enlarged in the race of competition that what was a good education formally for a man to make his way with in the world, is now only a very second-rate affair."[1]

Peto's education, however, would serve him well enough – both in his outstanding feats of construction and his unusually enlightened understanding of the enormous workforce he employed. At the age of 14, he was apprenticed to his uncle, the respected builder Henry Peto. For the next seven years, he worked alongside the tradesmen during the day before learning about architecture and the theory of construction in the evening. By the end of his apprenticeship, he was entrusted with site supervision and – according to his son – could lay 800 bricks a day.[2]

His son, the author of a 'memorial sketch' in his memory, quotes a commentator of 1851 who noted that: "He worked… not as the relative and future heir of one of the leading contractors of the kingdom, but as if he was destined during his lifetime to earn his livelihood

as a journeyman… and there cannot be a question that, besides the inestimable utility he derived from the insight thus voluntarily acquired into the mechanism of labour so essential to his calculations in its employment in vast organised masses, he also thus became familiarised with what may be called the idiosyncrasy of the English mechanic, and he has thus become enabled to convert such knowledge to the accomplishment of the moral results observable in his works."[3]

In other words, the insights Peto gained on the tools gave him both an understanding of the practicalities and logistics of immense projects, and an abiding empathy with the workers. This concern with their welfare would remain throughout his career.

Just as Samuel came of age in {1830}, his uncle died, leaving the business to him and another nephew, Thomas Grissell. In their early days in business together, they put in the lowest tender (by £400) for Hungerford market, but the Earl of Devon, Chairman of the Committee, asked Samuel to withdraw on account of his youth. According to his son, the young Peto told them that "if they would wait he would fetch his partner, who looked old enough for anything; adding that if his juvenile appearance was so much against him he must take to wearing spectacles."[4]

The cousins won the contract and built on the reputation already established by their uncle earlier in the century, when the firm had worked on the London Coliseum. They went on to win several other important projects, including the rebuilding of the prison at Clerkenwell. Based on the new Pentonville jail, this was a 'model prison' and an important part of the changing attitudes towards the treatment of prisoners. Although the new system would not stay in vogue for long – its emphasis on silence and segregation had a detrimental effect on the mental health of too many prisoners – it was significant because it marked the beginning of a new emphasis on improving, rather than just punishing, criminals. Built at a cost of £28,000 "on the model of Pentonville but in a rougher and less expensive style of architecture"[5], contemporary accounts stress that the demolition of the previous edifice formed a substantial part of the project.

Around this time, the cousins also won contracts for the Reform Club and the Lyceum Theatre, and on 6th April {1840} submitted the lowest tender for the construction of a new London landmark, Nelson's Column. Their price of £17,860 beat their nearest competitor by a mere £80. "The pillar is to be 50 feet higher than the Duke of York's column, and the figure of Nelson will be without a cloak," it was reported.[6]

1   Morton-Peto, H. 1893. *Sir Morton-Peto: A memorial sketch, printed for private circulation.* London: Elliot Stock p. 4
2   Ibid. pp. 6–7
3   Ibid. p. 10
4   Ibid. p. 12
5   Hepworth Dixon, W. 1850. *The London Prisons: With an Account of the More Distinguished Persons who Have Been Confined in Them.* London: Jackson and Walford p. 229
6   Urban, S. 1840. 'Domestic Occurrences', in *The Gentleman's Magazine.* Volume XIII June p. 642

{1830}
Altes Museum in Berlin, designed by Karl Friedrich Schinkel, begun in 1823, is completed.

{1840}
Bristol Temple Meads railway station opens in England, designed by Isambard Kingdom Brunel.

{1842}
Construction of Berry Hill, near Halifax, Virginia starts.

{1846}
Trinity Church in New York City, New York, United States is completed and consecrated.

The column was being funded by public subscription, which raised £17000, and reports in contemporary magazines reflect the public interest in the project. Despite this, the Memorial Committee ran out of money during the construction. Amid fears that the half-built column represented a threat to public safety, and a concern about "damage, particularly to the new National Gallery, that may be done if it fell", the government was obliged to step in with funding.[7]

As well as winning and constructing one of the most prestigious projects of the day, Grissell and Peto took a keen interest in new building technologies. During their work on this great London landmark, Grissell won a Telford Medal in silver from the Institution of Civil Engineers (ICE) for his paper "Description and Model of the Scaffolding used in erecting the Nelson Column".[8]

1840 also saw them commence work on the new Houses of Parliament. The old palace at Westminster had been destroyed in a fire six years previously, so the landmark we know today was bequeathed to us by the early Victorians. The cousins won major building contracts "for the range of buildings fronting the river from Westminster bridge to Abingdon street, the speaker's residence and the libraries being included. The second contract was for the houses of lords and commons, the Victoria Hall, Great Central Hall, Royal Gallery and House of Commons Office. The third was for St Stephen's Hall and porch."[9]

According to the records at the ICE, they were also early adopters of a novel technology for hauling bricks up scaffolding. The report remarks on the efficiency and beneficial effect on the safety of labourers of this "brick-raising machine", reporting that "Mr Grissell approved of the machine, and had found it very serviceable at the Houses of Parliament.[10]"

They were becoming extremely well-known. A contemporary limerick ridiculing liars ends with the claim that the reprobates "raised stories faster than Grissell and Peto"[11], a pun that suggests the builders had become household names.

However, the partnership between the cousins was to cease during the building of the parliament buildings. They had already worked successfully under Brunel on the Great Western Railway line from Hanwell to Langley, and this had given Peto a taste for the high-risk, high-reward world of railway building. Grissell, being more in the mould of his risk-averse contemporary William Cubitt, preferred to stay with more predictable contracts. The cousins decided to work apart, although the split was of a most amicable nature.

"Between the cousins there was perfect accord," Peto later wrote. "It is a great source of satisfaction to me to reflect that during the six and a half years we… never had an unpleasant word or misunderstanding. I cannot speak too highly of him either in our connections of business or friendship."[12]

Indeed, after their joint business accounts were closed and it was found that Peto was owed an extra five thousand pounds, he wouldn't hear of having the accounts re-opened. Grissell never forgot this, and would leave the same sum in his will to Peto's son.[13]

Peto's personal life took a sad turn in early {1842}, when his wife Mary, the sister of Thomas Grissell, died weeks after childbirth. Maternal mortality rates were still high at that time, and Mary's health had never been excellent. Peto published a private memorial in her memory, noting that "Frequent indispositions, and constant delicacy of constitution, had often excited many fears in the minds of anxious relatives that their cherished flower might fade long ere the shades of evening closed around it."[14]

Now a widower with four children, Peto married Sarah Ainsworth the following year. She was the daughter of a prominent Baptist, and Peto converted to this denomination himself. Indeed, Peto's second marriage seems to mark the beginning of a new phase in his career, which would become as notable for public service as it was for building.[15]

He became treasurer of the Baptist Missionary Society in {1846}, a post he would hold for 21 years. In this capacity, he financed five missions, including the liquidation of a debt of £9000 incurred in a mission to Jamaica. He also sent the Jamaican mission a temporary structure for use as chapel, and the surrounding area is still known as Mount Peto.[16]

7   See Brooks, E. C. *Sir Samuel Morton-Peto Bt, 1809–1889: Eminent Victorian, Railway Entrepreneur, County Squire, MP. Bury*: Bury Clerical Society p. 35

8   Newton, W. *The London journal of arts and sciences (and repertory of patent inventions)* VXXV. London: W Newton, at the Office of Patents p. 342. See also Anon. 1844. *Year-book of facts in Science and Arts: exhibiting the most important improvements and discoveries of the past year*. London: David Bogue p. 73

9   Morton-Peto, H. 1893. *Sir Morton-Peto.* p. 15

10 Minutes of the Proceedings of the Institution of Civil Engineers 1844 Volume 3 p. 222

11 Dickens, C et al. 1863. *Bentley's Miscellany*. Volume XIV. London: Chapman and Hall p. 242.

12 Morton-Peto, H. 1893. *Sir Morton-Peto.* p. 10

13 Ibid. p. 17

14 Morton-Peto, S. 1842. *Divine Support in Death, A few memorials Or The life and death of a beloved Wife, for private circulation in her family and amongst her personal friends*. London: Clowes p. 23

15 Chown, J. L. 1943. *Sir Samuel Morton-Peto, Bart… The man who built the Houses of Parliament*. London: Carey Press p. 5

16 Ibid. pp. 7-8

Nelson's Column | London, United Kingdom

{1852}

House of Commons in
the Palace of Westminster
designed by Charles Barry
and August Pugin is
completed.

{1854}

St. George's Hall in
Liverpool, England is
completed.

{1855}

Victoria Tower in London
is completed.

Peto also became a parliamentary candidate for Norwich in 1847. Unfortunately, navvies working for him on the eastern counties railway attacked a rival candidate. According to his son's memoir, "Their aggressiveness cost Mr Peto £70. Two hundred, after regaling themselves freely, lost self-control and attacked Norwich men, armed with bludgeons." To control the situation, "the navvies were decoyed into a train with no means of return until after the election".17

At the time, the Truck System, whereby labourers were paid in tokens that had to be spent at extortionate site shops, was common practice. The navvies and their families suffered enormous poverty as a result, and became notorious for alleviating their suffering through monumental drinking sprees and riotous behaviour. For this, they were almost universally despised, but Peto had the humanity to put their behaviour in the context of their conditions. He said of the Irish labourer, "I know from personal experience that if you pay him well and show him you care for him, he is the most faithful creature in existence; but if you find him working for 4d a day, and that paid in potatoes and meal, can we wonder that the results are as we find them?"18

With this in mind, his treatment of his own workers was exemplary by the standards of the time. As he embarked on his parliamentary career, he spared no exertion to improve the conditions of all ordinary workers on the railways. Speaking out against abuse of the navvies, he said: "I shall get the ill-will of every contractor, but for this I care nothing. I am serving the poor fellows, and promoting the cause of morality."19

This commitment to welfare was widely recognised and applauded. No less a person than the Bishop of Norwich was reported to have said "Mr Peto was a Dissenter, and he envied the sect to which he belonged the possession of such a man; he would gladly purchase him at his own price, and heartily he prayed that he would erelong become a member of the Church of England… All down the line he had met with his agents, and had found them not merely giving direction and instruction, but also giving to the men religious and school books for the education of themselves and their children, and thus showing them that education can civilise the mind, reform the habits and elevate the understanding. The gin shops were left deserted, and the schools were full."20

The business grew rapidly and Peto quickly won the respect of the great railway engineer Robert Stephenson. With his new partner Edward Betts (who married his sister Ann), Peto was soon working on massive railway projects all over the world. The rewards were high, but in 1847 he found himself in a tenuous position due to the massive liabilities he incurred.

In 1847 he wrote, "I now have £200,000 owing to me, and get it I cannot, but I trust my way will be made clear without sacrifice; but it must be sometime before the clouds clear away, and it has been very anxious work – these things come perfectly unexpectedly, and are not to be guarded against in large affairs."21

The problem was that his debtors were not always as forthcoming with their payments as he was himself. It was not the last time that he would find himself vulnerable to cashflow problems, but on this occasion he managed to retrieve the situation.

His confidence is evident in a letter to his wife the following year: "I shall, as I am doing, pay off obligations I incurred from the want of good faith of those who are indebted to me, and from the gain in this and other works have much larger surplus than will cover any loss, and I am sure it is my duty to you and the children to do this."22

By the middle of the century, he was involved in a number of diverse projects: the Buenos Aires Great Southern Railway; Algeria's first railway; the grand trunk railway of Canada; and the Victoria Bridge at Montréal. Also, his work on the Royal Danish railway was crucial to the country's butter exports and earned him the lasting appreciation of the King of Denmark.

The importance of Peto's railways to the development of national economies around the world also gained him private audiences with influential people. For example, Peto met the American President of the time (Andrew Johnson, Lincoln's successor) and found him to be refreshingly down-to-earth, while his meeting with the Emperor and Empress of France led to the following comment by the Emperor, who was delighted with the railway in Algiers, "I hope you will enjoy the consciousness that you are an instrument in advancing the civilisation and happiness of my people."23

Despite this success, Peto continued to be active in parliamentary affairs. Standing for re-election in Norwich in {1852}, he and his fellow candidate for the Liberals, Edward Warner, took a stand against bribery. In the political system of the time, bribery was an entirely feasible way of winning votes, since the franchise was so restricted. Only men who met the requisite property qualifications were entitled to cast a vote, a system that excluded most working men, as well as all women. Despite this, Peto and Warner held numerous public meetings to gather support.

17 Morton-Peto, H. 1893. *Sir Morton-Peto.* p. 56
18 Ibid. p. 57
19 Ibid. p. 82
20 Francis, J. 1851. *A history of the English railway: its social relations and revelations: 1820–1845.* London: Longman, Brown, Green and Longmans p. 271

21 Morton-Peto, H. 1893. *Sir Morton-Peto.* p. 20
22 Ibid. p. 21
23 Ibid. p. 43

Their ordinary followers, who could not vote themselves, undertook to patrol the streets of Norwich to ensure that no underhand transactions took place. The Norwich Mercury noted the "determination of the non-voters to keep watch and ward to prevent bribery."24

Through their efforts, both Peto and Warner were returned to parliament, unseating the eldest son of the Duke of Wellington. At the close of the election, the triumphant Peto said, "I rejoice beyond expression to say, that not one shilling has been spent in contravention of the law. The election has been conducted on independence and purity of principle."25 This was so remarkable that it was suggested by one commentator that it might be "the first time, probably, a Norwich election has been conducted with purity". In an age which took its religious convictions with extreme seriousness, the example of conviction triumphing over corruption was seen as both a religious and a political victory. "The city has been redeemed, and the people have done it!" rejoiced the Norfolk News.26

Peto's commitment to his anti-bribery stance was such that he later relinquished the opportunity to build the Portuguese railway system, rather than provide what were believed to be the necessary bribes to the decision-makers.27

Ever a friend to the ordinary people, Peto became an ardent advocate of the {1854} Payment of Wages Bill to stop the Truck System. Although an accepted practice at the time, Peto found it abhorrent. He is quoted as saying that

"from twenty-five years experience he could conceive of no reason why there should be a departure from the rule that a man's wages should be paid in the current coin of the realm… he never paid wages other than in money, and always took care that the men had it in sufficient time to derive the full benefit of it for their families."28

Equally as important, he provided the impetus to construct the first military railway. British troops were beleaguered and starving in the Crimea, and the situation was becoming a national disgrace. Peto was certain that a railway could provide the British with the strategic advantage they needed, so he convinced the Duke of Newcastle to approve the building of the Crimean railway. Then, he resigned his parliamentary seat in order to accept the government contract with the necessary propriety. He and his partners then made an undertaking to build it at cost.29

The future General Gordon, then a young lieutenant, wrote: "The civil engineers of the railways have arrived, and we hope soon to see the navvies and the plant. No relief that could be named will be equal to the relief afforded by a railway. Without the railroad, I do not see how we can bring up guns and ammunition in sufficient quantities to silence the guns of the enemy."30

Speed was of the essence, and seven miles of track were laid in as many weeks in early {1855}. They cut corners where necessary, as the line was intended to be temporary, and supplies were soon travelling up the line. In addition, the sick were being evacuated to the hospital ships bound for Scutari and the heavy guns that could go by rail were being transported. This allowed the British to make a decisive contribution to the war, with the Russians being particularly fearful of their bombardments.

In the words of Cooke, a historian of the period, "The war at the beginning had been fought as the Napoleonic Wars had been fought… The end was settled by bloody and overwhelming artillery bludgeoning made infinitely easier by the railway… In a few short months, Morton-Peto and Beatty [the engineer] and their men had enabled the art of war to be taken from Waterloo to the Somme."31

There is, of course, a darker side to any improvement in the efficiency of warfare, and heavy casualties resulted from this advancement. Nevertheless, Peto's actions saved British troops from unbearable suffering in the freezing Crimean winters, and enabled the evacuation of wounded soldiers from the front.

Interestingly, it also did nothing but good for the ordinary navvies and their perception in the eyes of the British public. Their efforts had been crucial in turning around the course of the war and improving the welfare of the troops. To quote Cooke once more, "The once-feared navvy had suddenly become a hero." 32

24 Reported in, *The Eclectic Review 1852,* Volume IV. London: Ward and Co. p. 257
25 Ibid.
26 Ibid. p. 256
27 Morton-Peto, H. 1893. *Sir Morton-Peto.* pp. 37–38
28 Ibid. p. 61

29 Ibid. p. 31
30 Ibid. p. 32
31 Cooke. B. 1997. *Grand Crimean Central Railway: the railway that won a war.* Knutsford: Cavalier House pp. 150–1
32 Ibid. p. 31

Houses of Parliament | London, United Kingdom

{1859}

Big Ben in London becomes fully operational.

{1861}

Arlington Street Church in Boston, Massachusetts, United States is completed.

{1866}

The building of Nationalgalerie starts (Berlin), designed by Friedrich August Stüler and Johann Heinrich Strack.

{1872}

Albert Memorial in London, designed by Sir George Gilbert Scott, is opened.

Peto gained a baronetcy, thus becoming Sir Morton-Peto in recognition of his services to the Crown. With his government contract finished, he also stood for parliament once more, and in {1859} was returned as MP for Finsbury.

His massive business interest in the construction of international railways never prevented him from taking an intelligent interest in politics and national affairs. For example, in {1861} he tried – and failed – to get a bill passed that would allow dissenters from the Church of England to be buried in churchyards. As a Dissenter himself, he objected to being classed with suicides and the excommunicated. He argued that a universal belief in The Bible meant there was no reason to deny a Dissenter burial within the Church.

Peto's position was both practical and humane, but in an age when theological differences were a foremost concern, his attempt provoked a flurry of angry rebuttals – some still preserved in pamphlet form – and the Bill was defeated.

He also advocated the building of iron, rather than wooden, ships "and argued that by contracting with builders of unquestioned reputation, for the construction of iron hulls at fixed and certain prices, there would be much economy, and that the present costly establishments in our dockyards might be greatly and permanently reduced."[33] His views on the defence policies of the day, and their cost to the taxpayer, were farsighted but out of step with the views of the majority; Peto was once again the subject of the indignant pens of Victorian pamphleteers.

Now well into late middle-age, Peto took on one last – disastrous – contract with Betts. They took shares instead of money for the Metropolitan extension of the London, Chatham and Dover Railway. Peto's own liability – effectively giving him control of the railway – was huge, and he was relying on borrowings from one of the foremost banks of the day, Overend, Gurney and Co. Unfortunately for Peto, this bank (historically a cautious Quaker concern –the celebrated Quaker prison reformer Elizabeth Fry came from the Gurney family) had taken to high-risk speculations promising massive returns. After these profits failed to materialise, the bank suspended payments. It was May {1866} and many firms were bankrupted in the fallout, including Peto's. As in his previous financial crisis, he was solvent on paper; owed more by his debtors than he owed to his creditors. Nevertheless, he was discharged as technically bankrupt.

Peto's affairs were subjected to a thorough investigation, but once more his character remained unblemished. As one correspondent wrote to him in a spirit of congratulation, "after a most searching and hostile inquisition, the imputations which have for two years beclouded your good name have been dispelled and the wound has not left a scar. The most violent of your assailants, the Daily Telegraph, is now among the foremost to express satisfaction at the result of the investigation."[34]

He had a similarly comforting letter from an eminent preacher, the Reverend Spurgeon, who wrote that "A little time ago I thought of writing to condole you in the late tempests, but I feel there is far more reason to congratulate you than to sympathise. I have been all over England in all sorts of society, and I have never heard a work spoken concerning you in connection with late affairs but such as showed profound esteem and unshaken confidence I do not believe that this ever could have been said of any other man placed in similar circumstances."[35]

Nevertheless, Peto had to resign from his seat in Parliament and from the Society of Builders, and could no longer compete in business on the scale he had once enjoyed. Talk of work on the Danube and in Russia came to nothing, and a lack of confidence was believed to have lost him the work. When Betts died in {1872}, Morton undertook his last job, the construction of the Cornwall Minerals Railway.

He retired to Eastcote, Pinner, where he continued to serve the wider community as a Justice of the Peace and Deputy Lieutenant. He spent his final years in Tunbridge Wells.

Although the final years of his career were not a success in worldly terms, Peto could look back on a lifetime of achievements and energetic exertions on behalf of the less fortunate. In addition to his tireless commitment to the welfare of his workers, he contributed generously to asylums and orphan schools. Also, he had been pivotal in ensuring the success of the Great Exhibition of 1851 by providing a guarantee of half a million pounds.

He was also remembered for small kindnesses, such as the story told by one colleague to Peto's son. While working with Peto on the American railways, the colleague had been amazed to find that, on hearing he had no pocket photo album of family pictures such as the one Peto himself treasured, Peto had immediately dispatched one to his colleague's wife with instructions to send some keepsakes for the long contract abroad.[36]

When Peto died at the age of eighty, the funeral address was given by Dr Angus, the principal of Regent's College (one of the many worthy institutions to benefit from Peto's generosity). Dr Angus was able to comment on Peto's devout character, telling the congregation that he "had known Sir Morton for nearly fifty years amid prosperity and adversity. He found it hard to say in which condition he admired him most, for he had been faithful in both."[37]

33 Morton-Peto, H. 1893. *Sir Morton-Peto.* p. 67

34 Morton-Peto, H. 1893. *Sir Morton-Peto.* p. 48
35 Ibid. p. 53
36 Ibid. p. 110
37 Chown, J.L. 1943. *Sir Samuel Morton-Peto.* p. 14

Houses of Parliament | London, United Kingdom

{1865}

George Gilbert Scott wins the competition to design St Pancreas railway station in London.

{1867}

Queen Victoria lays the foundation stone for the Royal Albert Hall.

Tower Bridge | London, United Kingdom

Sir Herbert Henry Bartlett

{1842–1921}

President (1892–93) of the Institute of Builders

Sir Herbert Bartlett was born in the village of Hardington Mandeville, Somerset. At the age of 15, he set off for London to pursue his ambitions and initially gained an apprenticeship with an architect, and then with a civil engineer.

In {1865}, at the age of 23, he joined a building firm run by John Perry, a successful East End carpenter who had built his business from scratch.

"Little is known of the early life and work of Mr John Perry," says the 1907 Perry and Co history, "but it is safe to assume that his lot was a hard one."[1]

Rather like Thomas Cubitt, Perry seems to have believed in the direct employment and close supervision of tradesmen, and he was able to employ more assistants and acquire larger premises at a steady rate.

After starting in carpentry, Perry had gradually added other trades to the business. By the 1860s, when Sir Herbert joined the firm, Perry and Co was involved in everything from gas works to bridges and railway stations. This breadth of involvement only increased over the years, as the company history indicates, "It is difficult to name any class of building in the construction of which Messrs. Perry and Co. have not had considerable experience. Their work ranges from the cottage of a country labourer to the palace of a foreign potentate; from mission hall to cathedral; from workhouse to palatial hotels."[2]

However, perhaps the firm's most important contracts were for hospitals. The company not only secured the contract for St Thomas's Hospital, worth £350,000, in {1867}, but also worked later on Great Ormond Street and the London Hospital.

St Thomas's was designed by Henry Currey FRIBA, an exponent of the pavilion principle, which tried to improve medical outcomes at a time when 'bad air' was blamed for disease. By using architecture to create hospitals with the maximum levels of ventilation and light, it was thought that infection would be reduced. Florence Nightingale, who later established a training school for nurses there, was enthusiastic about the design.[3]

Therefore, St Thomas's was as much a symbol of modernity and progress as the gas works and railways built by Perry and Co. The firm also constructed many of London's great warehouses, in places such as the East India and Victoria Docks, and landed many significant government maintenance contracts.

In 1872, after seven years with the firm, Sir Herbert became a partner alongside Perry's sons, as the founder looked to retire. Sadly, all of the sons died at a relatively young age, leaving Sir Herbert as sole proprietor by 1888. However, this did not harm the company's fortunes. On the contrary, it was reported that Sir Herbert had a "remarkably clear head for figures. He had no Shareholders, no Board of Directors. He costed everything himself, put in a tender, did the job, and, if he'd done his sums properly – and he usually had – made a profit."[4]

This assessment comes from a memoir published by his grandson Sir Basil, a professional writer urged by fellow-writer Beverley Nicholls to put his reminiscences in print. However, Sir Basil believed that monuments such as Tower Bridge were "lamentable buildings, beloved only by Sir John Betjeman, which have somehow survived the bombings and the demolition squads."[5]

Kenneth Clark (the great historian of culture and taste) observed that, in the 1920s, the architecture of the nineteenth century was widely considered insufferable.[6] Men like Sir Basil formed their tastes at this time and often never revised their opinions. What's more, the Herbert Bartlett that Sir Basil remembers is a distant and rather frightening figure, extremely solitary in his habits and given to the kind of eccentricities so closely bound up with our image of the Victorian patriarch.

Certainly, Sir Herbert was retiring enough to have little appetite for the civic honours he might have acquired through his business success. His activities outside of business – his funding of the Shackleton expedition to the South Pole, for example – were usually enterprises that he could undertake without attendance at a formal dinner. In fact, his donations to University College London were initially anonymous, and intended to endow a department of Eugenics – although this was not destined to become an abiding area of scientific interest.

1 Anon. c1907 *A Record of the Works of Perry and Co. Builders and Contractors* p. 2

2 Ibid. p. 3

3 Cook, G. C. 2002. *Henry Currey FRIBA (1820–1900): leading Victorian hospital architect, and early exponent of the "pavilion principle"*. Postgraduate Medical Journal 78. pp. 352–359

4 Bartlett, B. *Jam Tomorrow*. Elek. p. 107–108

5 Bartlett, B. Ibid. p. 18

6 Clark, K. *The Gothic Revival 1968*. Murray: London p. 2

Tower Bridge | London, United Kingdom

{1874}

The Opéra Garnier, designed by Charles Garnier is completed.

{1920}

The Cenotaph in London, designed by Edwin Lutyens, is completed.

{1920}

The Legislative Building in Winnipeg, Manitoba, Canada is completed.

However, in {1920}, Sir Herbert did accept credit for his generosity via the creation of the Bartlett School of Architecture, which remains one of the most distinguished built environment departments in the UK. This reflects his interest in science and progress, but he was also concerned with other forms of charitable giving. A poster from the Great War, held in the Imperial War Museum, lists him as Honorary Treasurer for the Soldiers and Sailors Dental Aid Fund, and it seems likely that many other instances of his charitable assistance are now unrecorded.

His wife Ada Barr, whom he married in {1874}, was an equally retiring soul. Remembered by her grandson as timid and probably frightened of the servants[7], she was dutiful in helping with the accounts and running the house, but most unlikely to push Sir Herbert into more public engagements. They lived with their nine children in an enormous house in Cornwell Gardens, London, which their son Hardington described as "dirty, inconvenient and old-fashioned, and furnished entirely with foreclosed mortgages."[8]

According to Sir Basil, his grandfather was such a solitary person that he even eschewed dining in the company of his family: "Whenever he arrived back from the office, he expected to be served at once, not on a tray in his study, but in the dining room. No matter how many people were dining there, the table had to be cleared, while he ate alone, surrounded by papers."[9]

Similarly, he travelled in a separate train compartment from the rest of his family, whom he abandoned to the third class carriage, and took shooting holidays alone or with friends. However, he did allow his family to enjoy his other pastime, a steam-yacht called The Gladys.

His grandson remembers these occasions as far more relaxed: "Strangely enough, the atmosphere on board the Gladys was quite different from the atmosphere at Cornwell Gardens. And I suspect it must have been different, too, at my grandfather's works at Bow, because no amount of bullying could have made workmen work as his men did for him."[10]

It appears that Sir Herbert was extremely conscientious about the welfare of his men. This cannot be ascribed to any religious sense of duty (it was recorded that "He built numerous churches, but entered them only for an occasional wedding or funeral."[11]), but seems rather to have sprung from the diligence that also characterised his approach to work.

By the standards of the day, the health and safety aspects of the construction of Tower Bridge, for example, were exemplary. In the words of the project engineer, "It is gratifying to note that the loss of human life during the construction of the bridge has not, considering the magnitude and the nature of the works, been great. In all, seven men have met with fatal accidents, and at least one of those was the result of sudden illness, or of a fit."[12]

As the company's men constructed the masonry of Tower Bridge, the firm was also increasing its involvement with the railways – tunnelling under the Thames to create the Bakerloo line and many of its most famous stations. This project, then known as the 'Baker Street and Waterloo Tube Railway', ran into a fault in the London clay as they burrowed under the Thames, meaning that compressed air was needed to keep water out of the tunnel.

"Extra large compressing engines were laid down so that a greater amount of fresh air was supplied to the men than in any previous operation of this character," says the company history, which also states that the firm retained a doctor to monitor the health of the men.

As well as helping the welfare of the workforce, the adoption of technology for improving the air had a positive impact on the speed of construction. This forward-looking approach to problem-solving later led to the firm winning the contract to rebuild the Cwm Cerwyn Tunnel, a feature of the Port Talbot railway that appeared dangerously close to collapse. Winning such contracts suggests that Perry and Co was one of the leading construction firms on tunnelling projects, but that was only one element of the company's order book.

While Tower Bridge must be the firm's best-known project, it also created landmarks such as the Piccadilly Hotel (now the Meridian, Piccadilly), which was designed by Shaw, and Mile End's People's Palace. This impressive complex, where Queen Victoria herself laid the foundation stone, brought facilities for learning and leisure to thousands in the East End.

7 Bartlett, B. p. 69

8 Ibid. p. 60

9 Ibid. p. 63

10 Ibid. p. 60

11 Ed, Jeremy. D. J. 1984. *Dictionary of business biography: a biographical dictionary of business leaders active in Britain in the period 1860–1980.* London: Butterworths p. 205

12 Barry JW in Welch, C. 1894 *Short Account Of The Tower Bridge* p. 57

"Among Mr Bartlett's treasured possessions are the mallet and plumb rule with which Her Majesty set the stone," the company history records.13

This would not be the company's only task with a royal connection. For Queen Victoria's Diamond Jubilee, the firm created the barricades for public safety, a task they performed again when the troops returned from the Boer War. In addition, the company built temporary stands for the coronation of Edward VII and other notable events.

When the company history was compiled, Perry and Co was adapting to the technologies of a new century. "The firm have ever been to the fore in following up new branches of industry… with the advent of the all conquering motor vehicle, the firm immediately took in hand and constructed two of the first garages for the new motorbuses; the largest garage built for electric broughams, and one of the first for petrol cabs."14

However, despite this commitment to technological advance in business, Sir Herbert never personally embraced the advent of the motor car, travelling to work in a horse and carriage until the end of his life.

According to Sir Basil, "He didn't believe in showmanship. The whole empire was run from a smallish office in Victoria Street, with the help of Mr Sharp, his secretary. All the detailed planning was done in the works at Bow."15

After the First World War, Sir Herbert stepped down and handed over control to Sir Basil's father, Hardington Bartlett. Soon after, the King of Belgium asked the firm to help rebuild the town of Dinant, which had been destroyed during the hostilities. Hardington departed to catch the ferry to Ostend, but then disappeared – seemingly, washed overboard although some suspected foul play. Could a business rival have been involved? The truth was never established, but the effect on Sir Herbert and on Perry and Co was decisive.

According to Sir Basil, "The shock of my father's death broke my grandfather. All the fizz went out of him, and he became an old man. Within the year he was dead himself. He realised, of course, that without my father his empire would collapse. And sure enough… by 1926 the proud old firm was bankrupt."16

Before he died, Sir Herbert had tied up a huge amount of money in trust for his wife. According to Beverley Nicholls, "By comparison with the saga of the Bartlett family inheritance, the plot of Bleak House seems of almost childish simplicity."17

The evidence suggests that the family connection with building did not quite die with the old firm. In 1935, when Prunier opened a restaurant in St James, they chose a building erected some years before by Perry and Co. The contractor for the alterations is listed as the Bartlett Trust Ltd18 and it seems likely that Sir Herbert's prudence and foresight allowed an interest in the building and property business to outlast the sad demise of the firm.

13 Anon. c1907. *A Record of the Works of Perry and Co. Builders and Contractors.* p4
14 Ibid. p. 6
15 Bartlett, B. p. 108
16 Ibid. p. 122

17 Ibid. p. 10
18 *'St. James's Street, West Side, Existing Buildings', Survey of London: volumes 29 and 30: St James Westminster, Part 1 (1960)*, pp. 472–486. www.british-history.ac.uk/report.aspx?compid=40623 Date accessed: 03 January 2009.

{1818}

William Strickland (architect) starts to build the Second Bank of the United States in Philadelphia.

{1837}

Great Conservatory, Chatsworth, designed by Joseph Paxton becomes the largest glass building in the world (demolished in 1923).

{1852}

King's Cross railway station in London is completed.

{1859}

Red House in Bexleyheath, England is designed by Philip Webb and William Morris.

Church of St Bartholomew the Great | Smithfield, United Kingdom

Frederick John Dove

{1830–1923}

President (1893–94) of the Institute of Builders

Frederick John Dove was a member of one of the great family building businesses. His grandfather was a village carpenter in Sunbury on Thames, and his father (William) moved to Islington in the 1820s to start building up the business in earnest.[1]

Islington was still an out-of-town staging post for coaches at this time, but it was growing rapidly. In about 1827, William Dove completed a contract for 50 houses in the area, and was also working on small commissions for the local Church Wardens by the 1830s.[2] Civic buildings and ecclesiastical commissions would be an extremely significant source of work for the firm as the century progressed, but its first major commission was the Islington Literary and Scientific Society's premises at Almeida Street, completed in {1837}.[3]

After three of William's sons, the youngest being Frederick, joined the business, the formal partnership of Dove Brothers was instituted under William's guidance. It was {1852}.

The business was solid, but Dove Brothers was not quite in the league of the Cubitts or Morton-Peto. Whereas Cubitts employed thousands, the 1861 census reported that Dove Brothers had 210 men and 11 boys on the books.[4] The business was steady rather than spectacular – as Summerson notes, "No fortunes were made… and the head of the firm lived modestly at the yard in Islington."[5]

Even so, the firm was the most prolific of the church builders – by 1900, it had completed no less than 130, of which more than fifty had been built before William's death in {1859}.[6]

Church-building took place on an unprecedented scale in the nineteenth century. Urbanisation had created areas where there were many times more parishioners than seats in the local churches. The fear of a revolution (such as the one that had taken place in France) cast a long shadow over the nineteenth century in England. The ruling classes believed that thousands of working people were being removed from the civilizing influence of Christianity – and this elite did not like the fact. Church-building had begun in earnest with the Church Building Act of {1818}, which had resulted in 214 "Commissioners' Churches", as they were known.[7] This did not wholly satisfy the perceived need for new churches, however, and the building and restoration of places of worship carried on throughout the middle years of the nineteenth century. As well as the church's own subscriptions, affluent private individuals also donated substantial sums for church-building. The need for new churches, and debates about the right way to build them, remained high on the agenda for decades.

As a leader in the field of ecclesiastical building, Dove Brothers worked with some of the greatest architects of the day, including Sir Gilbert Scott. A glance at the balance sheet, however, shows that this success did not always translate into cash. It is reported that "with an overall trading profit in 1866 of £2,650, less than half of the church contracts had proved useful, contributing puny sums rarely exceeding £100."[8] Even so, specialisation seems to have paid off in the end – the firm's profits tripled the following year, a level of success maintained throughout the 1860s.[9]

Church-building, and the fascination with gothic revival architecture that fuelled it, was tapering off by the latter years of the century. Although Frederick was the youngest of the three brothers, the company history reports that it was his decision to encourage the partnership to diversify. "When the church building boom subsided, the gap was more than filled by commercial developments in and around the City of London: there followed Central Hall, Westminster, and the prestigious Australia House."[10]

1 Braithwaite, D. 1981. *Building in the Blood: The story of Dove Brothers of Islington*. London: Godfrey Cave Associates p. 7

2 Ibid. p. 20

3 From: 'Islington: Economic history', *A History of the County of Middlesex*: Volume 8: Islington and Stoke Newington parishes (1985), pp. 69–76. www.british-history.ac.uk/report.aspx?compid=999 Date accessed: 24 February 2009.

4 Summerson, J. 1973. *The London Building World of the Eighteen Sixties*. London: Thames and Hudson. p. 12

5 Ibid. p. 15

6 Braithwaite, D. 1981. *Building in the Blood*. p. 26

7 Clarke, K. 1962. *The Gothic Revival: an essay in the history of taste*. London: John Murray p. 95

8 Braithwaite, D. 1981. *Building in the Blood*. p. 29

9 Ibid.

10 Ibid. p. 12

{1893}

The refinery for Pacific
Coast Borax Company
is the first reinforced
concrete building in
the United States.

{1893}

St. Mary's Cathedral
in Glasgow, Scotland is
completed.

{1893}

Broad Street Station –
Philadelphia, designed by
Frank Furness is enlarged. At
the time, the largest passenger
railroad terminal in the world.

A particularly important contract was for the foundations of GE Street's new Law Courts. Many familiar names are on the tender list – the brothers were bidding against some of the most successful firms in London – and the most expensive of the twenty tenders was for £68,347. George Myers bid £57,435 and Henry Lee, the first president of the Builders' Society, was just fractionally cheaper at £56,500. William Cubitt bid £46,555, while the canny Herbert Bartlett was, perhaps, quietly confident that the tender submitted by his firm (Perry and Co) would be competitive yet solidly profitable at £44,973. But even this price was shattered by the Dove Brothers' tender for £36,755, the lowest by an appreciable margin. (This information comes from an old tender book now stored in the local history archives of Islington library.)

Perhaps unsurprisingly, the brothers lost money on the project. Braithwaite observes that, since it was their largest project to date, it "was commendable that the overall losses were contained to £1323."[11] Besides, it paid off over the longer term, by announcing the firm's arrival into a bigger league of contracting. Other large projects would follow.

One of the most significant steps of this later phase in Frederick's career was the firm's move to central London. Dove Brothers took lease of a yard from the Mercers' Company, at what was then the highest ground rent in the capital.[12] This city office still yielded the company a profit in time, since the firm was able to rent out part of the building and attract an increasing number of commercial commissions, including a bank in Fleet Street and more factories and warehouses.

However, this did not mean that the brothers gave up their traditional church-building work. Their reputation continued in this area, and it's interesting to note that when the Christian Science Church arrived from America at the beginning of the twentieth century, it wanted Dove Brothers to build its first London church.[13]

The firm also continued to do a significant amount of ecclesiastical restoration work. One of the most significant contracts of this kind was for the church of St Bartholomew the Great at Smithfield. Founded in the 12th century by a courtier of Henry I, it is one of London's oldest churches. The repair it needed by the late nineteenth century attracted the attention of some prominent London citizens. When the firm's involvement in the project first commenced in the 1860s, the restoration was under the chairmanship of Builders' Society founder-member William Cubitt. And the second restoration committee was presided over

by the Archbishop of Canterbury! The architect for this second, more extensive restoration was Aston Webb, later President of both the Royal Institute of British Architects (RIBA) and the Royal Academy. Dove Brothers' work as the contractor won Webb's most fulsome approval. On the 19th November 1886, he wrote:

"Gentlemen, I have to report that in accordance with the instructions received from your Executive Committee, the work approved by you at the East End has been completed by Messrs. Dove Brothers within the contract time and to my entire satisfaction… The work has been carried out without the removal of a single worked stone from its original position, and it is certainly to the credit of the Contractors, the Clerk of Works and the Workmen, that in spite of the erection of very heavy scaffolding throughout the Church for the re-roofing, the whole has been set up and removed without any injury to the fabric, so that it is now handed back to the authorities with the ancient work in every way in as good a condition as it was received by us."[14]

Clearly, the firm's expertise in dealing with ancient churches allowed for a standard of workmanship – and a very modern sensibility regarding historic structures – that impressed even the most discerning of clients.

Aston Webb also drew up the plans for some of the buildings for Imperial College, which was expanding in the early twentieth century. Designed "in a restrained Tudor domestic style", the eastern side was specially modified to include large and extensive windows to accommodate the study of botany as well as plant pathology and physiology. The Dove Brothers erected this part of the building between 1912 and 1914, at a total cost of £16,698.[15]

Frederick took an active part in the affairs of the Institute, and served as president in {1893}. In his presidential photograph, Frederick has cultivated an extensive beard, which he reputedly told his grandson "obviated the need to wear a tie".[16]

A founder member of the incorporated body, Frederick took a keen interest in those issues occupying the most insightful minds within the Institute. In the paper he presented to an IOB meeting

---

11 Ibid. p. 36
12 Ibid. p. 42
13 Ibid. p. 44

14 *Appendix: Restoration committee and architect's reports (1863 and 1885)', The records of St. Bartholomew's priory [and] St. Bartholomew the Great, West Smithfield: volume 2* (1921), pp. 536–553. www.british-history.ac.uk/report.aspx?compid=51798 Date accessed: 24 February 2009.
15 From: 'Imperial College', *Survey of London: volume 38: South Kensington Museums Area* (1975), pp. 233–247. www.british-history.ac.uk/report.aspx?compid=47532 Date accessed: 24 February 2009.
16 Braithwaite, D. 1981. *Building in the Blood.* p. 13

of 1888, The Architectural Profession and the Building Trade, in relation to each other, critically examined, he looked at the relationship between architect and builder, noting that:

"It cannot but have struck most of us who have had a moderate experience of our business that, for a long time past, say twenty years, practical knowledge has been becoming less and less a part of an architect's education, and, in the rising generation, it promises to become extinct."[17]

His paper is not just a part of the sometimes difficult dialogue between the two professional interests. It also displays an awareness of the need for the Institute to act as the mouthpiece for builders' views on issues – such as the parliamentary debate on the registration of architects, which could have affected the working conditions of builders:

"At a time when there is every reason to suppose that, ere long, an application will be made to Parliament by the Royal Institute of British Architects, or by some other body in connection with the architectural profession, for power to give certificates of qualification to those seeking to practice as architects, and who, without such certificate, would be unable to follow that vocation; at such a time, it appears to me, not out of place for members of the building trade – more especially those who are members of this Institute – to examine the relations at present existing, and to offer an opinion on the subject from a builder's point of view."[18]

As well as taking an interest in the drafting of legislation, the Institute also kept an eye on the case law that was likely to impact upon builders' interests. Frederick appears to have been one of a number of builders asked to act as an expert witness in a pivotal case. It is stated in the Institute's annual report for 1887 that "A case of "Error in Quantities" having been brought to the notice of the Institute involving principle of great importance to builders, the Council, under the advice of their solicitor, have requested Mr Stanley G. Bird, Mr. John T. Chappell, Mr J. Howard Colls and Mr F J Dove to accede to the request of the builders interested, and give evidence on their behalf."[19]

Twenty years later, in 1907, two events gave Frederick John enormous pride. One was being formally presented to King Edward VII, following the firm's successful completion of the contract to build University College School in Hampstead, and the second was his son, Frederick Lionel, beginning his presidency of the Institute.

When Frederick John died in 1923, he was commemorated as a founder of the IOB. He had also been the last-living founder member of the London Master Builders Federation. In the words of the family firm's historian, "In the span of just over 70 years, the range of buildings for which he was at least nominally responsible was remarkable by any standards."[20] At the time of his death, the City Press described Frederick as 'the father of the building trade'. Not only had he successfully grown one of London's most respected building firms, but he had also worked tirelessly to ensure that the building profession gained the representation it deserved through his involvement with the Institute and other industry bodies.

Frederick Lionel built on the success of his father, who had retired at the beginning of the First World War, and personally superintended the urgent repairs to St Paul's Cathedral. And as the century progressed, the firm continued to work on prestigious projects, such as the repair of Westminster Abbey.

Frederick Lionel reached milestones such as the presidency of the IOB much earlier in life than his father, and expanded the firm far further, but it was Frederick John who had the vision to move the family business from Islington to where it could take its place amongst the most respected contractors in London.

---

17 *Proceedings of the Institute of Builders 1888.* Volume 1 Part 3 pp. 49–50
18 Ibid. p. 49
19 Ibid. pp. 46–7

20 Braithwaite, D. 1981. *Building in the Blood.* p. 7

The Cenotaph | London, United Kingdom

Sir Edwin Lutyens

{1869–1944}

Founder, Architecture and Surveying Institute

Sir Edwin (Ned) Lutyens had eminent artistic influences from the start of his life. At the time of his birth, Ned's father was studying under Landseer and the baby was baptised Edwin Landseer Lutyens in his honour.

When Ned grew older, he studied at the South Kensington School of Art before being articled to the architectural practice George and Peto. The junior partner, Harold Peto, was the son of Sir Samuel Morton-Peto, and their practice was eminent in the field of late Victorian domestic design. Ned, however, didn't feel he was learning, and went into practice on his own after six months.

Although he was not yet twenty, pivotal influences were already informing Lutyens' style. A great admirer of William Morris and the Arts and Crafts movement, Ned would be strongly drawn to natural materials and a kind of unpretentious, humane vernacular throughout his working life.

Before long, he was introduced to the garden designer Gertrude Jekyll, which was the beginning of a long-standing collaboration. Their meeting was fortuitous for them both – Miss Jekyll was far older than Lutyens and had failing eyesight, so garden design became the substitute for her outlet as a painter. Similar in their respect for vernacular craftsmanship, they almost immediately began to work together, quickly winning commissions.1 In 1896, Jekyll entrusted Lutyens with the design of her own house, Munstead Wood.

Munstead Wood's design featured many components that would become integral to Lutyens' later work – gables, tall chimneys and low casement windows. It has been noted that Lutyens tended towards gardens that were "formal and extremely complicated"2, continuing his stonework out into walls, steps, terraces and pillars. Jekyll provided the rambling roses and vines to soften the scene with an impression of careless, natural beauty and to create the illusion of a semi-tended rural cottage garden that threaded around Lutyens' carefully thought out structures.

Formality, even grandiosity, had been the hallmark of houses and gardens in the preceding Victorian age. Lutyens could work with both classical conventions and a charming English style. His undoubted mastery of form and proportion enabled him to create many (still sought-after) country houses combining elegance with an impression of honest rusticity.

To a generation looking for comfort and homeliness rather than grandeur, he was the perfect architect; his reputation grew steadily. Princess Louise gave him a small commission to advise on some alterations. Lutyens fantasised about great royal commissions, which did not materialise, but the brief patronage from a member of the royal family undoubtedly did his career no harm.3

Another admirer of his work was Edward Hudson, founder of Country Life Magazine and owner of Lindisfarne castle. Lutyens remodelled the castle to provide comfortable living quarters, while Jekyll transformed the gardens and included upturned boats as novel garden sheds4 – an unusual interpretation of the Arts and Crafts predilection for using local materials! Lutyens also built one of the last English castles – Castle Drogo – for a rich grocer.

The 'Englishness' of Ned's architecture – particularly in his numerous Surrey country houses – was recognised and celebrated by contemporaries as a nostalgic imagining of the past. Lutyens' England is an elegant, comfortable and humane rural idyll that was never actualised for the majority. Nevertheless, at a time when romanticising the past was in vogue, his work struck a chord with the rising middle class.

Although derided by some as a 'society' architect, he became representative of something more far-reaching than individual house design. He was, rather, rapidly becoming the architect who best symbolised the epitome of English domesticity. He designed the scenery for two plays by his friend J. M. Barrie – one of these, "Peter Pan", included a faithful recreation of the night nursery used by Lutyens' children5. In 1912 he designed street scenes for a "Shakespeare's England" exhibition at Earl's Court, an event that reflected the contemporary vogue for celebrating the nation's historic achievements6.

1   Scott-James, A. and Lancester, O., 2004. *The Pleasure Garden*. Frances London: Lincoln Ltd. p. 98

2   Ibid. p. 102

3   Martin, B and Sparke P. 2003. *Women's Places: Architecture and Design 1860–1960*. London: Routledge.

4   Plantagenet, S. F. 2007. *Castles, Scotland and the English Borders*. Newton Abbot: David and Charles. p. 111

5   Howard J. 2005. *Shakespeare Reproduced: The Text in History and Ideology*. London: Routledge

6   Searle, G. 2004. *A New England, Peace and War, 1886–1918*. Oxford University Press.

{1930}

40 Wall Street (then called the Bank of Manhattan Trust Building) is completed and is briefly the tallest building in the world.

{1930}

The Chrysler Building (New York), designed by William van Alen, succeeds the Bank of Manhattan Trust Building as the world's tallest building.

{1930}

Chachmei Lublin Yeshiva building in Lublin, Poland is completed.

{1931}

Commerce Court North is completed in Toronto, Canada and becomes the tallest building in the British Empire.

With his genius for design recognised and his work achieving cultural resonance, Lutyens' talents were soon required for more momentous commissions. In 1912 he was elected to the Delhi Planning Commission, and began designs for the Viceroy's House. Perhaps coincidentally, he was married by this time and the son-in-law of a previous Viceroy of India. Following George V's announcement the previous December that the capital of India was to be moved, a massive project would be undertaken to create a new Imperial capital. The following year, Lutyens was appointed joint architect for New Delhi with Herbert Baker.

Baker and Lutyens had first met years before, when both worked for George and Peto[7]. Baker had gone on to become a successful architect in the British Empire, notably with the design of the Union Building in Pretoria, South Africa. The relationship was cordial initially, but they would fall out in an incident that Lutyens would later call his "Bakerloo".

Lutyens wanted to build the Viceroy's House on the peak of a small but steep hill. However, it was felt that this might give out the wrong political message, and that the administrative Secretariats – whose building Baker was designing – should be placed at the same height as the symbol of Imperial rule. It is reported that "to please his colleague… Lutyens agreed to place the Viceroy's House at the back – or west end – of the hill… on the condition that the great road leading to it between the twin Government offices would be so gently sloped that a visitor approaching the House from the Great Palace would see the whole dome and the entrance portico down to the bases of its columns.[8]"

Undoubtedly, Lutyens thought he was offering a fair and simple solution, but the realisation of the plan did not achieve the necessary gradient for success. When viewed from the side of the Great Palace, Lutyens' design is almost entirely obscured by the road, with just the tip of the dome coming into view over the horizon. "You are faced by a wide stretch of asphalt rising to the sky and hemmed in by sharply diminishing red stone cliffs, instead of a broad vision of palatial majesty linking yet commanding the introductory buildings.[9]"

This problem in the execution of Lutyens' vision should not, however, be allowed to detract from the overall accomplishment. The palatial building, the surrounding gardens and extensive town planning (which the critic David Watkins likens to an English garden city[10]) were a momentous achievement for an architect with negligible formal training. Although the buildings were completed in {1931}, only 15 years before the end of British rule in India, they remain an iconic contribution to imperial architecture.

Long before the completion of works at Delhi, however, Ned received a commission of far more lasting significance. Lutyens had already served as one of the principal architects on the Imperial War Graves Commission and, in 1919, Lloyd George informed him there was a requirement for a catafalque (temporary construction to hold a coffin) for the peace celebrations. It was to be a non-denominational edifice, to mark the contribution of all of the fallen, and it was needed in a fortnight.

The Whitehall Catafalque would have been commissioned on the spot, had Lutyens not immediately suggested that what was really required was a cenotaph: a monument to someone buried elsewhere. The new name was agreed and according to Hussey, Lutyens' biographer, "The commissioning, conception and rough through finite design of the Cenotaph took place within six hours; probably less."[11]

Lutyens was a patriot, a man of his time who would have viewed the war as a necessary evil. Nevertheless, when he travelled to France, he was palpably moved by the scale of the loss that had taken place. He wrote to his wife:

"The cemeteries – the dotted graves – are the most pathetic things, especially when one thinks of how things are run and problems treated at home. What humanity can endure, suffer is beyond belief…"[12]

While he shared in the zeitgeist of patriotism, his religious beliefs were most unusual for his generation – he had a Universalist view of spiritual truths, which made him disinclined to base his work on any specific iconography. It is noticeable that Lutyens' memorial is the essence of simplicity, and particularly that it lacks any reference to Christian symbolism.

7 Reilly, C. 1931. *Representative British Architects*. Jeremy Mills Publishing: Huddersfield p. 47

8 Butler, A. *The Lutyens memorial –the architecture of Sir Edwin Lutyens*. London: Country Life Limited Vol 2 p. 29

9 Ibid.

10 Watkin, D. 2005. *A History of Western Architecture*. London: Lawrence King Publishing.

11 Hussey, C. 1953. *The Life of Sir Edwin Lutyens*. London: Country Life Limited. p. 392

12 Lutyens to his wife, 12th July 1917. In Percy. C. and Ridley. J. (Eds.) 1995. *The letters of Edwin Lutyens to his wife Lady Emily*. London: Collins pp. 349–50

He took a similarly straightforward approach to the design of the Monument to the Missing of the Battle of the Somme, which stands at Thiepval. "Just as in the case of the Cenotaph", says one commentator, "Lutyens brilliantly managed to convey an embodiment of nothingness, an abstract space unique amongst memorials of the Great War."13

Considering the sensitivities of the Establishment, it seems surprising that more protest was not mooted. However, the Anglicans had long been inclined to distance themselves from anything overly ornate and, besides, these memorials were a response to an unprecedented enormity of grief and loss. The sober austerity of Lutyens' work matched the need of the hour. As the other European powers endeavoured to repatriate their dead, Britain decided not to attempt this dreadful task. On street corners across Britain, empty coffins marked with the names of the dead were strewn with flowers.14 Lutyens' empty tomb was Everyman's resting place, and its design does not exclude anyone through the trappings of faith.

His design was so well received that a permanent version of the monument was completed in 1920. If there had been any doubt previously, he was now established as a renowned architect.

Even so, one of Lutyens' greatest designs, for Liverpool Cathedral, was never built in full – only the crypt was constructed. He is remembered instead for prosperous country houses and for the classical city buildings of his later period, as well as his designs for London housing schemes, where he always tried to bring light to depressing locales.

Lutyens was also sole consultant to the engineer Bressey on the 1937 Highways Development Survey that "envisaged a new capital, with ringed roads round it, arterial roads crossing it, tunnels under the parks, and the reconstruction of narrow streets"15. The advent of World War Two would take attention away from street planning for some years, but much of the southern portion of the M25 follows the Bressey/Lutyens route for a 'southern orbital'.

Lutyens' stature ensured that he received numerous awards and recognition for his achievements. He was knighted just after the First World War and, in {1930}, his work at New Delhi earned him the title Knight Commander of the Order of the Indian Empire. Towards the end of his life, he was awarded the Order of Merit, becoming part of a very select chivalric order devoted to recognition of intellectual, scientific and artistic achievement.

There was some debate in Whitehall on whether the Order of Merit should be given to Lutyens or his younger rival Giles Gilbert Scott. When the opinion of Kenneth Clark (the eminent art historian) was sought, he replied: "Lutyens, without a moment's hesitation… even now, at the nadir of his fortunes, while his style is out of date without yet being historical, he overtowers all rivals."16

This was in 1937, five years before Lutyens received his award – and many more before the great architect would become 'historical' and receive the approbation of posterity. Hussey notes that Lutyens would be criticised by younger architects for a lack of interest in such issues as economics and social planning. These were the foundations he took for granted, while concerning himself with his "absolute values – beauty, truth and human dignity".17

The most striking facet of any brief survey of Lutyens' work is his sheer versatility. From rustic idylls to Indian palaces, from stage sets to the London orbital, he approached his work with humanity and humour. Lutyens was rooted in tradition, yet able to create monuments that transcended culture and belief; his sheer versatility, as well as his innate sense of form and proportion, underline the exceptional nature of his talents.

13 Winter, J. 1998. *Sites of Memory, Sites of Mourning: The Great War in European Cultural History.* Cambridge: University Press

14 Worpole. K. 2003. *Last Landscapes: The Architecture of the Cemetery in the West.* London: Reaktion Books

15 Minney, R. J. Hore Belisha. L. 1960. *The Private Papers of Hore-Belisha.* London: Collins. p. 80

16 Martin, Stanley. 2006. *The Order of Merit I.B.* London: Tauris. p. 190

17 Hussey, C. 1953. *The Life of Sir Edwin Lutyens.* London: Country Life Limited. p. 586

Sir Manuel Hornibrook OBE

{1893–1970}

Honorary Fellow (1969) of the Institute of Building

The "Father of the Australian Building Industry" was a man of large stature, both literally and metaphorically.

6'2" and seventeen stone, he is remembered as a man of limitless energy, great vision and boundless optimism. Sir Manuel was a native of Queensland, although his parents were both originally from Cork. His father had "all the qualities of a pioneer"[1] and used determination and persuasion to bring about the building of the first road in Obi Obi district. Unfortunately, Sir Manuel's father died at the young age of thirty seven, leaving a widow and seven children.

Never much interested in school work, Sir Manuel, the second eldest boy, proved a quick learner when it came to the economic realities of the family's new situation. He immediately got himself a paper-round, and then apprenticed to a local builder at the age of 13.

His boss's son remembered Sir Manuel as having a remarkable memory, with the ability to glance at a plan just once and then finish the job without further reference to it.[2] By his late teens, Sir Manuel was pricing jobs and at 19 took responsibility for a building contract to build 19 houses at Kangaroo Point.

Even at this early age, Sir Manuel felt ready to go into business with a partner. However, the partnership was short-lived and Manuel headed home again. On his arrival at the train station, Sir Manuel got chatting to the station mistress, who needed a house building. So at the age of 21, he landed his first solo contract to build a house, labour only, in his home town.

The business grew and in {1918} he expanded into civil engineering. In 1925, he commenced construction of his first major bridge, a reinforced concrete structure for the Queensland Main Roads Board. He would go on to build over 100 bridges in his lifetime, the work for which he is best remembered.

In {1927}, he won the commission to build the William Jolly Bridge in South Brisbane. To do this, he devised the 'sand island' method of pier construction.[3] The city council's consulting engineer considered it unconventional, and insisted that each set of river pier foundations be completed prior to any progress payment. The method worked. Instead of floating precast cylinders into place, artificial islands were built over the pier sites, forming an enclosed area with steel sheet piling filled with sand. The cylinders were then lowered into position (by open-dredging through the sand island and river bed) and then embedded into the rock and sealed with concrete by workmen using airlocks.

Sir Manuel employed a medical officer on site, and all workers were issued with special cards to ensure speedy and effective treatment if they should suffer from the 'bends'. When one unfortunate worker fell into the river, injuring his spine, Sir Manuel dived in to save him and then provided extra financial compensation to the family. There were no fatalities, and the accident record was extremely impressive at that time for a job of such complexity.

After that, Sir Manuel's firm was much in demand for similar projects, including a reinforced concrete bridge across the Coomera River. The company also constructed five bridges, as well as many miles of road, in Maroochy Shire. However, as the 1930s wore on, there was less public money available for this kind of project: a problem for a major contractor with a large number of people to keep in work.

However, according to his biographer, "Manuel Hornibrook's reaction to circumstances such as these was typical and a pointer to his attitude to adversity when it came his way throughout life. He never conceded defeat and would always seek to find a way out of a difficulty."[4]

It has been noted that one of Sir Manuel's bridges is a contender for the title of the first-ever BOT (Build-Operate-Transfer) project.[5] This style of procurement is now common, but would have been virtually without precedent at the time. In fact, new legislation, The Tolls on Privately Constructed Road Traffic Facilities Act of {1931}, had to be passed by the Queensland government in order to give the idea a framework.

Nevertheless, the Hornibrook Highway – Australia's largest bridge – was built during the Great Depression using private finance. It opened as a toll bridge, charging one shilling, in October {1935}. Attracting funders would have been no mean feat given the dire economic conditions, but the bridge was eventually handed to the government, debt free, 40 years later.

At 8086 feet, it was one of the longest bridges ever built at the time, and the longest in Australia by some way. A new dual-carriageway bridge now carries any motorised traffic, although the old bridge – with its Art Deco portals at each end – is still used by cyclists and pedestrians.

In addition to building numerous bridges and tunnels, Sir Manuel's firm made a major contribution to utilities provision in Queensland, before extending the business into New South Wales. This would not be the limit of the company's territorial expansion, however.

In the late 1940s, Sir Manuel's firm started contracting in New Guinea, beginning with the Daugo Island aerodrome and then major projects such as a wharf at Port Moresby and the Rouna Hydro Electricity Scheme (a joint venture with an American firm). The company also built the Markham River Bridge, its first in the Territory, despite heavy flooding – a feat that only augmented Sir Manuel's growing reputation.

As well as working on other major bridges – e.g. the Maryland River Bridge, Northbridge, Iron Cove Bridge – the company won phase two of the Sydney Opera House project: the construction of the shells. There had been major problems for the contractor on phase one, but stage two, on the recommendation of Ove Arup himself, was awarded to Sir Manuel's company without recourse to tender.[6]

1   Browne, W. 1974. *A Man of Achievement*. PEP Enterprises: *Brisbane* p. 1
2   Ibid. p. 4
3   Ibid. p. 17
4   Ibid. p. 23
5   Morledge, R. Smith, A and Kashiwagi, D. T. 2006. *Building Procurement*. Oxford: Blackwell Publishing p. 206
6   Murray P. 2004. *The Saga of Sydney Opera House:* p. 42

{1918}
Police Headquarters are built in Copenhagen, Denmark. Designed by Hack Kampmann in a Neoclassical style.

{1927}
Shrine of Remembrance in Australia is completed.

{1931}
Empire State Building is completed and takes over as the tallest building in the world.

{1935}
The Hoover Dam (USA) is completed; at the time, the world's largest electric-power generating station and world's largest concrete structure.

The Minister for Public Works observed that "the shell structure of the roof, by its very nature, is a most difficult and complex work. Its intricate form and special constructional problems do not lend themselves to the preparation of plans and tenders in the normal manner. The engineers had the unique problem of devising the best and most efficient means of arranging for the erection of the shell superstructure. Work of a similar character has never before been attempted on this scale anywhere in the world."[7]

As an enlightened company such as Ove Arup had grasped, the nature of the innovation required full collaboration (sooner than normal) from all parties in the construction team – an early example of innovation shaping the contractual arrangements.

In the words of Sir Manuel's biographer, "This was the type of challenge to bring joy to Manuel Hornibrook's heart and he took an intense and active interest in every phase of the construction work."[8]

Adjustments to the supporting columns were required before construction of the roof could begin. These went down through the podium into the foundations, and had been constructed with the best available information at the time. But the design of the roof had since changed.[9] Work to demolish them from the half-built structure was slow – "it had taken two men with jack picks two weeks to cut away just one yard of the high-strength concrete."[10]

Sir Manuel therefore suggested explosives. This in itself took four months, with the explosions being carefully detonated at the height of the rush hour traffic – a bid to disguise the situation from journalists already aware of the time and cost issues facing the project. This ruse worked very well, until a lump of concrete landed on a passing ferry.[11] The subsequent headlines were predictably scathing.

Twenty columns then had to be strengthened in order to support the new roof design. The project was moving towards the realisation of the revised design, now recognised as one of the most iconic pieces of architecture in the world. But to observers at the time, it must have seemed just a highly expensive and chaotic project.

There were more than 2000 roof components[12], and Sir Manuel's team were making these pre-cast elements in situ, effectively setting up a factory on site. These sections could weigh up to 15 tons each and were held together by 350km of tensioned steel cable.[13]

At the age of 71, Sir Manuel climbed 150ft for a progress inspection. "He could climb to places where men years younger than he could not,

and indeed would not venture," remarks his biographer[14]. Due to the complexity of the design, Utzon, the architect, also insisted that Hornibrook should be hired without competitive tender for phase three.

Despite these successes, Sir Manuel had to deal with mixed fortunes in his later years. The flotation of his company was not as profitable as might have been hoped, taking place at a time of rising material costs. And, according to his biographer, the firm's involvement in a consortium to reconstruct the Mount Isa Railway caused a further strain on liquidity.[15]

However, Sir Manuel remained philosophical in the face of difficulty and continued to work not only for his company, but for the greater good of the industry. His lifelong commitment to professionalism would, in his later years, bring him several honours.

As a young man, he had joined the Queensland Master Builders – a relationship that would continue for the rest of his life. In 1926, he served as president of the Master Builders' Federation of Australia. This body identified the need for a professional institute, and Sir Manuel became only the second construction professional to serve as the president of the Australian Institute of Building. He worked to raise finance for the Institute's headquarters and invited the Prime Minister, Sir Robert Menzies, to open the new building in 1955. There, the Prime Minister presented him with the organisation's first Medal of Merit, for his contribution to the science and practice of building. One of those contributions was an annual travel grant to young construction professionals, which demonstrated his commitment to education and professional development.

He was knighted in 1960, and chose 'Nil Sine Labore' (nothing without work) for his personal motto. He retired in 1966 and busied himself with his last construction project – a doll's house for his granddaughters.

In 1969, the UK's Institute of Building made him an honorary fellow – an accolade that seemed to be the culmination of the many Life and Honorary memberships bestowed on him in his native Australia. His biographer describes it as the 'climax of his achievements'.[16] Sadly, Sir Manuel's health was too poor for him to make the journey to London to receive the fellowship. An IOB fellow therefore brought the scroll to Australia, hoping to present it to him at the annual dinner of the Australian Institute of Building, in Queensland. In the event, Sir Manuel was too ill even to attend a local event, so a friend received it on his behalf and a deputation brought it to his bedside.

Sir Manuel's biographer observes that many colleagues were delighted by this accolade, seen as the most prestigious in the Commonwealth, noting that it could be described in the same terms as those used by Sir Robert Menzies about the Medal of Merit: "awarded by the judgement of his peers, not by some incompetent tribunal… by those best qualified to know his worth."[17]

7 Browne, W. 1974. *A Man of Achievement*. Brisbane: PEP Enterprises. p. 90
8 Ibid.
9 Murray, P. 2004. *The Saga of Sydney Opera House*. p. 52
10 Ibid.
11 Ibid. p. 53
12 CBS Team. 2003. *The Romance of Construction* , India: CBS Forum.
13 Ibid.

14 Browne, W. 1974. *A Man of Achievement*. p. 90
15 Ibid. p. 108
16 Ibid. p. 144
17 Ibid. p. 147

William Jolly Bridge | South Brisbane, Australia

Sydney Opera House | Sydney, Australia

Sir Ove Arup CBE

{1895–1988}

Honorary Fellow (1976) of the Institute of Building

In the popular imagination, it is the architects, out of all the built environment professionals, who are thought to have the grand ideas about building and society. Builders and engineers, the public tends to assume, are the more prosaic, practical types who merely give form to the theoretical designs of others.

Many leaders challenge that idea, but perhaps no one illustrates the limitations of such generalisations better than Sir Ove Arup – theorist, pianist, trained philosopher and outstanding engineer.

In his youth, architecture was still seen as the dominant building discipline, the most fitting profession for 'the educated man'. However, according to an obituary for Sir Ove, he considered studying architecture but decided he would "rather be a good engineer than a second-rate architect."[1]

Undoubtedly, Sir Ove achieved his engineering ambition – without abandoning his training in philosophy. He constantly questioned the status quo and looked far beyond the parameters of practical problem-solving. In fact, his awareness about the impact of construction on society became more profound with the passing years. According to his son, "One could almost say that his life's work was a progression from the part to the whole. He started with the integration of design and construction, then came structure in architecture, the place of architecture in society, the impact of technology on society and ultimately the fate of society itself."[2]

Born in Newcastle, Sir Ove spent most of his childhood in Hamburg, where his father worked as a veterinary consultant for the Norwegian government. The young Ove studied philosophy at the University of Copenhagen, but did not find the subject wholly satisfactory.

As he recalled in a 1964 interview for the BBC[3], "In philosophy, there seemed to be no answers, only more and more subtle questions… I revolted against ideologies, philosophical systems, moral codes, on Kant's insistence on rectitude in preference to kindness."

As Europe was engulfed in the First World War, Sir Ove decided to train as an engineer. Introspective and somewhat agonised by religious doubts and an impassioned correspondence with a young woman called Olga (they resumed their correspondence at the very end of his life.[4]), his uncertainties made him gloomy. However, he concluded that "it wouldn't be the worst thing to happen, to become an engineer… irrigation plants in Mesopotamia, India, etc. A room, wine, a book an armchair, a pipe."[5]

While irrigation plants were not destined to become Sir Ove's speciality, he had secured a job with the leading firm Christiani and Neilson by 1922. Posted to Hamburg, the city of his childhood (and with German as his first language), Sir Ove lived in Germany as the economy entered a period of hyperinflation. Indeed, he took his salary straight to the bakers on payday, knowing that its value would have diminished by the next morning.[6] He became keen to leave, and was appointed to the company's London office.

In London, he met Ruth, a young Danish music teacher, whom he married in 1925. Ove missed the intellectual culture of the continent and started to apply his intelligence to the problems of the industry, publishing his first technical paper in 1926. Reflecting his life-long concern with breaking down disciplinary boundaries, the paper challenged a previous writer's notion that architects planned structures while engineers advised on the detail. He pointed out that this relationship could work just as easily in reverse, particularly on larger structures and in the light of new technologies and materials. In fact, Ove himself had a particular interest in the application of concrete, and his innovations with this material would be the basis of his most notable early projects.

In 1934, the UK's first Code of Practice for Concrete was published. Unfortunately, it gave little specific guidance to Ove about the achievement of the structural strength necessary for the projects he was engaged upon.[7] A year earlier, in 1933, he had calculated the requirements (from 'first principles') for building the Penguin Pool at London Zoo; an 'exercise in structural gymnastics' based

---

1   Arup, O. and Partners. 1988. *Sir Ove Arup 1895–1988*. p. 3
2   Ibid. p. 4

3   Quoted in Jones, P. 2006. *Ove Arup Master Builder*. New Haven: Yale University Press. p. 21
4   Ibid. pp. 13–14, 17–21, 44, 316–18
5   Ibid. p. 16
6   Ibid. p. 29
7   Sutherland, R. J. M. et. al. 2001. *Historic Concrete: background to appraisal*. London: Thomas Telford. p. 90

{1938}

The Lions Gate Bridge
in Vancouver, Canada
is completed.

{1963}

Bankside Power Station in
London, designed by Giles
Gilbert Scott is completed.

{1963}

Millbank Tower in London,
England is completed.

{1973}

Sears Tower in Chicago,
Illinois, United States,
becomes the tallest
building in the world.

on torsional strength.8 However, the most significant challenge of those pre-war years was Highpoint I, the block of flats he designed in collaboration with the architect Lubetkin.

This not only provided a platform for his talents, it caused Sir Ove to move jobs in order to work on the project: "Christiani and Neilson would not undertake the contract and Ove accepted the post of Chief Designer with J L Kier and Co in London… on the condition that Kiers would build Highpoint." 9

The design for this unprecedented structure dispensed with all beams and most columns, relying on reinforced walls instead. It was also remarkable for the movable wooden formwork Sir Ove designed to facilitate its construction.10

Sir Ove, in hindsight, saw many faults with the building. It was designed to be 'functional', but the cost of a watertight flat roof outstripped that of a traditional sloping roof. Looking back, he felt the structure was 'muddy', and there were faults with the service terminals. Furthermore, while the architect was promoting it as suitable for working class families, Sir Ove felt that "the working class are exactly the kind of people who would hate to live in these buildings with open planning". 11

This was just one of many robust debates between the two European multi-linguists, Lubetkin and Ove. The latter was interested in the Truth, which (despite his youthful rejection of the abstract) he sought within his work. According to Sir Ove, the left-wing Lubetkin was inclined to see all art as deliberately biased. While their partnership was not destined to be long-lasting or easy, in old age they each acknowledged a debt to the other's skills.12

In {1938}, Ove left Kier to go into business with his cousin, Arne Arup, and soon became embroiled in a controversy about air raid shelters. Through Lubetkin, Ove became consultant to Finsbury Borough Council, which was anxious to identify the best way, with limited resources, of protecting a largely working class population living in poorly constructed housing from the coming bombs. Sir Ove thought again of concrete, and his expertise in torsion. He designed large, spiral ramps (later known as deep shelters) with air conditioning and lavatories, which could provide large numbers of people with protection. They were also designed to be used as underground car parks after the war.

Although an elegant solution, it was not destined to meet with favour from the authorities. They felt that underground solutions – such as reopening the tube stations at night – could be bad for morale, and in any case favoured a theory of dispersion.13 They wanted to provide individual families with small shelters (such as the Morrison and Anderson) that offered protection unless there was a direct hit. The Ministry reasoned that since a direct hit was statistically unlikely, causalities would be kept to a minimum. Sir Ove found the arguments unconvincing, and engaged in long and increasingly urgent correspondence with the authorities to try to encourage a change of policy.

Ove's biographer records that eventually, in 1941, he was granted an audience with an anonymous official. This person agreed with Sir Ove's mathematical conclusions, but claimed that the priority was to give people confidence in the existing precautions to prevent panic. Sir Ove thought this a somewhat cynical view14 and later referred to the authorities' handling of the affair as 'deplorable'.15

A more satisfactory outlet for Ove's talents appeared during the preparations for D-Day. During the construction of the floating Mulberry Harbour, which was one of the building industry's most impressive contributions to the war effort, Sir Ove solved the problem of how to absorb the impact of boats as they landed. Improving on the existing technology (known as the 'Baker Principle', which used long and rigid fenders for protection), Sir Ove designed fenders that were loosely hinged, and so far better equipped to absorb the force of contact.

As Ove's biography explains, "By an ingenious but strictly calculated system of links and guides, forces were appropriately and sequentially displaced, although with one unintended side-effect: 'an unending screaming noise as they ground together along the side of the pontoon'."16

Sir Ove founded his consultancy in 1946, although it did not become known as Ove Arup and Partners until 1949. Through the Fifties, the firm set up offices across Africa and the UK and, in {1963}, in Sydney too. The company had been working on the Sydney Opera House since 1957, and it would remain involved until the building's official opening in {1973}. During the course of that project, the firm of 100 permanent staff had become a truly global presence, employing 1500 people.

8   Ibid.
9   Arup, O. and Partners. 1988. *Sir Ove Arup 1895–1988*. p. 3
10 MacDonald, A. J. 2001. *Structure and Architecture*. Oxford: Architectural Press.
11 Jones, P. 2006 *Ove Arup Master Builder* Yale University Press: New Haven. pp. 59–60
12 Ibid. p. 313

13 Due to the number of casualties inflicted, underground stations would eventually be adapted and opened as shelters.
14 Jones, P. 2006 *Ove Arup Master Builder*. p. 87
15 Ibid. p. 84
16 Ibid. p. 106

The Opera House, while a breathtaking technical achievement, took its toll on Ove. He argued bitterly with the architect, Utzon, who accused him of wanting control of the project: "How can a consulting structural engineer dare to encroach on the architect's work in such a fantastic damaging way?" 17

Utzon ultimately left the project, which was finished under Peter Hall, but the public rift was deeply hurtful to Ove: this consummate individualist was a life-long advocate for teamwork. According to Peter Jones, his biographer, the experience left him shaken – he never again worked so closely with an architect outside of Arup and Partners. He had always trusted the senior partner on a project, but now he doubted both himself and those around him – why hadn't he seen where the relationship was heading?

Nevertheless, Ave continued to preach holism, to promote a humanitarian and thoughtful ethos, and to strive for truth in his life and work. According to his son,

"he worked for a greater understanding between the professions associated with the creation of the built environment, but was never doctrinaire about how this was to be achieved."18

His Key Speech of 1970 remains the company's philosophy to this day19. According to this manifesto, work should be a source of pleasure and satisfaction. The other main tenet of his argument is that we should not seek happiness recklessly, at the expense of others. He believed we need to recognise that "no man is an island, that our lives are inextricably mixed up with those of our fellow human beings, and that there can be no real happiness in isolation."20

He also realised that principles are odifficult to live up to, and that a sense of proportion is a valuable gift. In order to reconcile this to his ideals, he suggested what he called the 'star system': that people should set their course towards a guiding star, or principle. Sometimes, deviation from this path may be necessary, but the star remains the navigational point to which the person should return once the difficulty is overcome.

Avoiding both absolutism and absolute relativism, this metaphor is an apt example of Sir Ove's unique perspective. As a philosopher, he never abandoned the exploration of moral problems. As an engineer, he had a gift for illustrating ideas in more concrete terms. In the words of his son, "For him the problem was never merely formal, but always real. Whether as engineer, philosopher, or simply as active friend, it wasn't just – what is the ideal building for this purpose? But also – how can we set about getting this built?"21

Sir Ove certainly built many impressive structures. His projects, too numerous to discuss, include the Kingsgate footbridge in Durham, Coventry Cathedral and the restoration of York Minster. However, his corporate legacy is equally impressive: a firm that continues to espouse his principles of liberal humanism. Both in terms of the concrete and of the philosophical, Sir Ove could build enduring structures.

17 Ibid. p. 232
18 Arup, O, and Partners. 1988. *Sir Ove Arup 1895–1988*. p. 4
19 Arup, O. *The Key Speech*. Available at **www.arup.com/_assets/_download/FAF05210-19BB-316E-40AFF213E48D1CF1.pdf** Accessed 6th January 2009
20 Ibid.

21 Arup, O, and Partners. 1988. *Sir Ove Arup 1895–1988*. p. 2

{1912}

Manchester Opera House in England opens on Boxing Day.

{1917}

Het Schip housing scheme designed by Michel de Klerk in Amsterdam is started.

Sir Maurice Laing
{1918–2008}

Honorary Fellow (1981) of the Chartered Institute of Building

Destined from birth for a career in construction, Sir Maurice Laing was a member of one of the great building families. The family business can be traced back to James Laing, himself the son of a stonemason, and the Cumbrian building firm he set up in the 1840s.[1]

By the turn of the twentieth century, James Laing's grandson, John William Laing, had inherited control of a growing contracting business based in Carlisle. In {1912}, the company landed a significant £20,000 contract to construct Carlisle's new post office. The firm's growing reputation for good work meant that by {1917}, it employed nearly 4,000 workers across various contracts, including some major installations for the war effort. When Sir Maurice was born in 1918, the company was building Crail aerodrome, a munitions factory and a firing range.[2]

After the war, the focus switched to housing – homes fit for heroes were now a national priority – and John Laing patented an 'Easiform' building system in 1924. This was described as "the proved system of shuttering and scaffolding for speedy house construction by unskilled labour and utilization of waste products".[3] Steel shuttering was used to form a mould, into which concrete could be poured to create walls. The fact that no skilled labour was required was a huge advantage, given that so many experienced men had lost their lives in the war.

John Laing was a committed Christian and did his best to be a benevolent employer, but his standards were exacting: those not measuring up could be fired on the spot.[4] When Sir Maurice first joined the business, his inexperience led to problems on a job he was supervising. John arrived to inspect and, appalled, delivered a lecture on his son's failings. According to the firm's historian, however, "Maurice was too much his father's son to take the dressing-down he received. 'You can talk to your men like that as much as you like,' he said explosively, 'but do it to me once more and you'll never see me again.' White with rage, John

Laing walked out of the site office and drove away. Fifteen minutes later, he returned and apologised. He never upbraided Maurice as caustically again."[5]

Sir Maurice had started work in the mid-1930s, when the Great Depression still affected building companies profoundly. He later said "I am grateful that I started work in 1935 because if I had gone to university I would have started at a later age and would never have seen the depression. It has influenced my whole life".[6]

According to Sir Maurice's nephew, Sir Martin Laing, "he was straight, quick tempered (but soon got over it) and ever a pessimist over the economy."

Like many of his generation, the experience of the great slump would never leave him. Sir Maurice's pessimism is much commented upon by those who knew him, and it seems hard to resist the conclusion that it stemmed from those difficult years in the 1930s. It is said he always saw trouble ahead too, although he could be encouraged out of this stance and his natural caution didn't stop him having a lively sense of humour.

"The first time he met my wife, he turned round and said 'You won't get rid of that one!'" says Christopher Laing, another nephew. "And he was right. I used to call him my wicked uncle – in the nicest possible way. He was always prone to teasing his family and had a very dry sense of humour. It was difficult to read him, because you didn't always know if he was joking."

(Sir Maurice spoke at an IOB conference in 1976, arguing against the mooted extension of direct-labour departments. He advised the minister present that if he wished to nationalise anything else, it should be crime "because we all want to make sure that that doesn't pay".[7])

There was always something of the firebrand about Sir Maurice, and this would manifest itself acutely as the Second World War took hold. The firm built 54 airfields during the war[8] and John Laing wanted his son to stay with the company and work on these projects. Sir Maurice, however, wanted to join the RAF and contribute to the war effort in the field of combat, rather than on the building site.

1 Richie, B. 1997. *The Good Builder. Laing PLC*. London: James and James
2 Ibid.
3 Ibid. p. 55
4 Ibid. p. 61

5 Richie, B. 2008. *Sir Maurice Laing*. The Independent. 26th February. Available at **www.independent.co.uk/news/obituaries/sir-maurice-laing-firebrand-head-of-the-laing-construction-company-who-became-the-first-president-of-the-cbi-787303.html** Accessed 9th October 2008
6 Goodman, G. 2008. *Sir Maurice Laing*. The Guardian, 25th Feb. Available at **www.guardian.co.uk/business/2008/feb/25/3?gusrc=rss&feed=business** Accessed 9th October 2008
7 Anon. 2008. *Sir Maurice Laing*. The Times 28th February Available at **www.timesonline.co.uk/tol/comment/obituaries/article3448584.ece** Accessed 9th October 2008
8 Goodman, G. 2008. *Sir Maurice Laing*. Accessed 9th October 2008.

{1964}
BT Tower in London,
England, is topped out
on July 15.

{1964}
Prudential Tower in
Boston, USA, is completed.

{1964}
Verrazano Narrows
Bridge opens in New
York Harbour.

It is reported that "John Laing persuaded the head of works department at the Air Ministry to stop his son's call up. A furious Maurice threatened to quit and risk imprisonment unless he could join."[9] He duly got his way and later distinguished himself by piloting only the second British glider carrying troops across the Rhine.[10] This was an incredibly dangerous mission – the casualty rate among glider pilots was 38% – and Sir Maurice was a relatively inexperienced pilot, but he managed to survive.

On his return to Britain and the family firm, Sir Maurice's influence increased rapidly. He persuaded his father to import new technologies from America, and to investigate the possibility of building in South Africa, where he had spent time as a trainee pilot. Alongside his brother Kirby, Sir Maurice took charge of what was now Laing PLC on his father's retirement.

Many notable contracts followed, including the construction of the M1. The company had been invited to bid for various sections of the road, but they ended up with all the contracts. As one writer notes, "It was Britain's first motorway, so there were no motorway specialists."[11] And due to the high profile of the project, the pressure was immense – and the weather terrible.

In November 1958, Sir Maurice sent the following message to his team via the company newsletter:

"We all know that the eyes of a large number of people in this country and in other parts of the world are upon the activities of those engaged on this contract. I should like to take this opportunity of congratulating them on all their marvelous achievements to date to wish them continued success. They are proving that it is possible to build 53 miles of motorway in 19 months, even when the first summer of the contract has been the worst in living memory."[12]

The road opened on time.

Laing also took on the contract for Coventry Cathedral, the first new cathedral in Britain for many centuries. Sir Maurice was a committed Christian, like his father, and he followed the firm's tradition for such buildings: it was built at cost. Sir Maurice's straight-talking approach eased the way in a somewhat difficult relationship with the architect, Sir Basil Spence, and the two became great friends. While Sir Maurice had managed to get Sir Basil to produce plans on time, it was clear that the steeple wouldn't be completed before the scaffolding was scheduled to be dismantled. Sir Maurice's solution? The scaffolding came down at the designated time, and the steeple was flown into position using a helicopter.

According to Christopher Laing, "His faith was overwhelming and affected all he did. It made him very determined and difficult to sway. He was trustworthy – his word was his bond. From my own point of view, he was a great mentor; I could always go to him to talk about difficulties. He was always happy to help."

As well as being involved in such landmark contracts, the company always took an active interest in research and development. During the post-war years of shortage, a Laing laboratory developed Thermalite – a cellular material made from pulverised fuel ash. It had clear potential, but needed refinement. Sir Maurice led successful negotiations to pay a royalty to Ytong (a Swedish firm that had already created a similar product), to share their expertise.[13]

This was not the limit of Sir Maurice's interests, however. In {1964}, he was the last president of the British Employers' Federation. Convinced that a more united representation of industry's views was needed for it to be effective, he became involved in the merger with the Federation of British Industries and the National Association of British Manufacturers. This new organisation was the Confederation of British Industry (CBI), and Sir Maurice was its first president. Helping to broker this merger was one of his proudest achievements. He was also a director for the Bank of England for 17 years, reflecting the great esteem in which he was held, both for his talents and his attitude to business.

Peter Harper, a former director of Laing PLC and Chairman of Sir Maurice's trust, first worked with Sir Maurice during the recession of the early eighties, a difficult time for businesses.

"He never cut corners when the going was tough," says Peter. "Even when redundancies were necessary, he did his best to look after people, paying them as much redundancy as he could afford, and stressing that it was the position, not the person, who was redundant. He had to take tough decisions about the structure of the business, but he did it with courage and compassion. He was concerned for the welfare of all, from the most senior executives to the blue collar workers. I remember him as someone who proved that you don't have to be a rat bag to be a good businessman."

9 Ibid.
10 Helme, D. 2005. *Sir Maurice Laing: a life in profile*. London and Carlisle: Laing.
11 Ibid.
12 Ibid. p. 57

13 Richie, B. 1997. *The Good Builder*. p. 122

His courage helped the company get through the difficult years of recession, and business subsequently picked up and remained very strong for the rest of the decade. Sir Maurice was committed to the continuation of his father's legacy of responsible leadership and a benevolent approach to business.

"He was genuinely interested in other people," says Christopher Laing. "When visiting projects, he talked to everyone, including the craftsmen and labourers. He was incredibly good at remembering faces and names. His attitude won him the respect of his peers in the industry. As a leader, he always wanted to make sure the workforce was looked after. His great relaxation was sailing, and he led his crewmembers with the same sense of responsibility and care. He would never knowingly lead anyone into danger."

As an offshore racer, he regularly beat fellow enthusiast Edward Heath in races in the 1970s and was a frequent competitor in the Fastnet race. He was also rear-commodore and trustee of the Royal Yacht Squadron and president of the Royal Yachting Association.[14]

This interest in sailing led to a donation of one million pounds to the Cowes Waterfront Project, where there is now an events centre named after him.

He also showed his generosity through the Maurice and Hilda Laing Charitable Trust, which he endowed for the relief of poverty and the spread of Christianity, and through his regular donations to other charities. Today, his Trust needs three full time members of staff, and is heavily involved in supporting church schools, evangelical television and various projects to relieve poverty.

Christopher Laing also remembers that Sir Maurice gave a lot to support natural health, which had done so much to help his wife Hilda. He helped to set up the Tyringham naturopathic clinic and established the Laing Chair in Complementary Medicine at Exeter University.

"He took the view that this should be available to the people, not just to the privileged" says Christopher. "This was typical of him – he always wanted the profits of business to go to good causes elsewhere."

In retirement, as well as maintaining an active interest in his charities and enjoying the opportunities for travel and holidays with Hilda, he retained his competitive spirit. In his mid-eighties, he was driving a Mini Cooper, and never quite lost the urge to make use of its rapid acceleration when alongside far younger drivers at the traffic lights.[15] The firebrand element of his personality, which had once led him over the Rhine, never truly died.

14 Anon. 2008. *Sir Maurice Laing*. Telegraph 19th March www.telegraph.co.uk/news/obituaries/1582124/Sir-Maurice-Laing.html Accessed 9th October 2008.

15 Helme, D. 2005. *Sir Maurice Laing: a life in profile*. London and Carlisle: Laing. p. 112

York Minster | York, United Kingdom

Sir Peter Shepherd CBE

{1916–1996}

President of the Institute of Building (1964–1965) and Honorary Fellow (1987)

Sir Peter Shepherd is remembered as a visionary who provided strategic leadership to the family business, The Chartered Institute of Building and the construction industry.

He combined rigorous thinking with courageous implementation, and these qualities played an essential role in the development of what was, for a time, the largest privately owned construction and manufacturing group in the UK. Furthermore, his presidency of the CIOB was pivotal to its modernisation and development, and has been equalled by few others in terms of its impact.

Sir Peter's son, Paul Shepherd PPCIOB, remembers that his father "always described himself as a builder by education, but an accountant by inclination." He also had a rare ability to predict the future. With these talents, he could have done many things, but banking would have perhaps been the most natural use of his abilities.

He was the third generation in the family business and his introduction to his construction career was a true baptism of fire. He had completed just one year at boarding school when the Depression of the 1930s caused major problems in the family building business of F. Shepherd and Sons. He had to leave school immediately (much to his delight, he later admitted) and started work in July 1931, at the age of 14.

His first day was spent working in the bedroom of his father (who was suffering from flu), helping to sort unpaid bills and to decide who they could afford to pay the next week. His father was more inclined towards the sites than to book-keeping, so Sir Peter's strong aptitude for accounting and administration enabled him to make a real contribution from the beginning.

These initial few months, which were brutal in commercial terms, were the defining period of his business career. Through that experience, he developed a life-long awareness of the need for strong and unrelenting financial control in any undertaking.

In 1947, his father suffered a stroke. Sir Peter was appointed Managing Director at the age of 31.

During the 1950s, his brother Donald developed the "Portasilo", a system for storing cement on site that greatly reduced waste. This was an important development, as materials were scarce during the early post-war years. It was very successful and led to the formation of Portasilo Ltd.

The combination of Sir Peter's business acumen and Donald's inventive skills was highly fortuitous. Donald had the ideas and Sir Peter found the finance to make them work. They combined with their other two brothers, Michael and Colin, to give a powerful impetus to the business.

The Shepherd Building Group was formed during the 1960s. F Shepherd and Son was superseded by this company and its thriving range of subsidiaries, which dealt with such specialisms as construction, development and concrete. Indeed, the subsidiary Concrete Services was one of the very first ready-mixed concrete companies. Portasilo and Portakabin were also important components of the Group.

All the companies progressed strongly through the 1960s, but at the end of the decade, business suffered when a major contract to build 4,000 homes using industrialised techniques strongly promoted by the Government went seriously wrong.

For Sir Peter, who had been so profoundly moulded by the task of steering a business through the Great Depression, it must have seemed as though history was repeating itself. Paul Shepherd remembers that for two years the company lived on a knife edge.

"I recall him saying to me that he did not think the company would survive," he says. "He advised me to get another job, since the importance of completing my engineering qualification was uppermost in his mind. But of course I didn't leave."

By that time, the company had grown to 7,000 employees. The organisation Sir Peter had to steer through these difficult times was, therefore, a very different business from the F. Shepherd and Sons of the 1930's. Nevertheless, yet again the business survived and life returned to normal around {1972}. Still, it had been perilously close. Sir Peter would be the first to acknowledge the total commitment and contribution of his three brothers and their extremely loyal staff to the company's survival, recognising that such talent and loyalty was amongst the greatest assets of a family business. Ultimately, however, Sir Peter's exceptional financial skill and leadership had been crucial to saving the organisation.

This experience further reinforced his belief that balance sheet strength was the prime need for a business. Paul recalls him saying that "I cannot envisage any amount of [company] cash that will make me feel entirely comfortable." Therefore, he resolved to "build a financial fortress".

At the same time, Shepherd Construction was carrying out the restoration of York Minster.

"The Minster was built in many phases," explains Paul. "The foundations of this massive building had been constructed using timber and the falling water table had caused the foundations to dry and destabilise."

The experts felt that the Minster could begin to collapse within 15 years. The solution was clear: a massive underpinning of the building, particularly the 65 metre high central tower, which weighed 16,000 tons. The big question was how this would be achieved.

It was a very challenging, five-year project. A major factor in its ultimate success was, long before the term became fashionable, a perfect example of 'partnering' between client, architect (Bernard Fielden), consultant (Arup) and contractor.

The design and construction was very much an iterative process. After trial excavations, the architect and Arup returned to the drawing board to design the next stage. This was followed by discussions with Shepherd Construction to decide whether the latest proposals were feasible.

During construction, the remains of the most northerly headquarters of the Roman Empire were discovered beneath the central tower. They were preserved in situ and later became part of an exhibition on the history of Christianity. When the work was finally completed, Sir Peter was always proud to say that, throughout the entire contract, the company had never delayed a single daily service.

A major commercial development during the 70s and 80s was the growth of Portakabin. Developed by Sir Peter's brother Donald, it rapidly became a household name. Again, the brothers' complementary skills combined on this project. Sir Peter managed the finances to fund the building of new factories, and this allowed the rapid development of this stream of the business. This manufacturing division complemented the more cyclical activities of the construction companies, contributing significantly to the later financial strength of the Group.

By the 1990s, Shepherd Building Group had grown to become one of the largest construction, manufacturing, engineering and property companies in Europe. It was Dun and Bradstreet "triple A" rated – Sir Peter had indeed achieved his goal of building a financial fortress.

The concern with financial robustness had clearly been shaped by Sir Peter's early experiences in the 1930s. This pre-war period had also moulded his views on the importance of education. He was always aware of the overwhelming need to improve teaching and broaden academic opportunities construction.

Sir Peter had perceived that construction in the 1930s often relied too much on crisis management. The problems causing the crisis were by no means always the fault of the contractor, but nevertheless the builder regularly came off worst in the ensuing arguments. Sir Peter realised that a lack of thorough education, and therefore an inability to articulate a case, commonly put the builder in a disadvantaged position within the construction team.

"In his early years on site my father observed these interactions", says Paul, "and questioned the tradition of deference. He felt it was totally wrong that the contractor took most of the risk, yet was not considered to be an equal member of the construction team."

During the 1930s, Sir Peter was travelling to Leeds to attend evening classes and studying for an HNC in building. The fact that only 110 students throughout the country completed the course in the year he qualified was clear evidence of the shortage of opportunities to study building before the war. Sir Peter quickly concluded that education was the key to giving builders equality with architects and other professionals.

{1972}

Brunswick Centre in
London, designed by
Patrick Hodgkinson
is completed.

{1972}

Catedral de Maringá in
Maringá, Paraná, Brazil
is completed and becomes
one of the tallest churches/
cathedrals in the world.

{1972}

Transamerica Pyramid in
San Francisco, California,
United States is completed.
It is the tallest skyscraper in
the San Francisco skyline.

{1996}

Oscar Niemeyer completes
the Niterói Contemporary
Art Museum in Brazil.

The experience that inspired this commitment to construction education was another early insight which remained with him for the rest of his life. It was his central belief that everyone had the potential to achieve, regardless of their background. During the 1960s, this conviction led him to pioneer the concept of the 'ladder of opportunity'. This remains enshrined in the philosophy of CIOB education to this day.

Given this deeply-felt commitment to education, it was indeed natural that he would become both a strong supporter and active member of what was then the IOB, and an insistent advocate for the development and improvement of the Institute.

The construction industry of the 1950s was struggling and chronically short of professionally-trained people. At the time, the Institute was perceived as a wealthy builders' club, chiefly notable for holding impressive dinners. Something had to be done.

"Father was captured by the mission and became something of a leader of the faction working for change," remembers Paul. "Fortunately, his vision was matched by a desire for change in Dennis Neale, who was then Institute Secretary. They became an exceptionally empathic and effective team. They also became great friends, working together for over 25 years to implement the radical changes that the Institute so desperately needed."

Paul is keen to stress that Sir Peter was not a lone voice for change. However, he was certainly a natural and cerebral leader who instinctively caught the mood and needs of the times. Perhaps more than any other contractor of his generation, his vision was crucial to the creation of enduring improvements in building education.

As well as becoming a tireless advocate for learning and scholarship, he did much to develop the local branch and centre structure of the Institute – much of which still operates today.

"This was a very bold move at the time, as the membership was not large enough to support such a structure," says Paul. "It was put in place in the anticipation (and the hope) of growth. To help grow membership, he then developed an innovative scheme encouraging every member to introduce a new member."

Leading fundamental change within professional institutions, which are, almost by definition, traditionally-minded bodies, inevitably involves both commitment and an ability to engage in and win robust debate. This is not a challenge for the faint-hearted, and must have taken enormous determination and energy, especially when one remembers the many other claims on the time of an industry leader. Sir Peter, however, was more than equal to the challenge, working tirelessly to champion the cause of modernisation against sometimes powerful opposition. His proposals for change ultimately won out, and the institute of today owes much to this achievement.

One of the most important innovations of his presidency was the change of name: the Institute of Builders became the Institute of Building. This change was later to become even more significant when the Institute applied for a Royal Charter several years later.

"He felt strongly that the name should reflect its status as a professional institution rather than a trade association," says Paul. "Father was a purist, and the need to develop an institute that was recognised as a credible and respected professional body was his absolute belief and mission."

Sir Peter was appointed Chairman of the Wool Training Board at the age of 48, an assignment which reflected his achievements in the field of education, and served in that position for several years. He was later asked to be chairman of the Construction Industry Training Board, which had been running with a large financial deficit and was not generally well-regarded by the industry. Paul recalls that two civil servants came to York to see his father, and announced that they were not leaving his office until he agreed to become chairman!

His achievements were recognised with the award of a CBE, and later a knighthood. He also received Honorary Doctorates from five universities, but perhaps the award which he appreciated the most was becoming an Honorary Fellow of the CIOB.

Sir Peter died in {1996}; ten months after he stepped down from the family business that he had served for 65 years.

Just as businesses in construction are commonly handed down the generations, the Shepherd family's involvement with the CIOB has spanned more than a lifetime. Paul had previously been Chairman of the National Contractors' Group, and was both a President of the Building Employers Confederation and the first Chairman of its successor organisation, the Construction Confederation. During the last conversation he ever had with his father, he was able to tell him he had been asked to stand for the junior vice-presidency of the CIOB.

{2000}

On May 12th the Tate Modern in London (a conversion of Bankside Power Station) by Herzog & de Meuron opens to the public.

{2000}

Emirates Towers in Dubai, United Arab Emirates are completed and open to the public.

While he regards this honour as partially a tribute to his father's achievements, it is undeniable that he has continued the family tradition by making a significant contribution to the Institute.

The only father and son to be made honorary fellows of the Institute, they have both given significant time and effort to improving the effectiveness of the CIOB and the wider standing of the profession. Sir Peter was the great educational innovator, and Paul has been pivotal in the development of CIOB International.

"When asked what my presidential initiative would be, I replied that 'my initiative' is not to have an initiative" says Paul. "I believed that there were already a number of good ideas in the pipeline which had not been fully implemented, and it was important to complete the best of the existing work rather than add to the number of half-finished projects.

"After my presidential tour in {2000}, however, I suggested we think again about how we worked overseas. That was how CIOB International began. To my mind, it was simply unacceptable to raise the expectations of overseas members with a presidential visit and then, perhaps only one month later, merely to hand over to your successor without creating proper continuity."

After he had left the family business, Paul gave up to 100 days each year to the Institute, working tirelessly to build the international membership. Collaborating closely with the CIOB's Deputy Chief Executive, Michael Brown, he helped the Institute to make huge strides, notably its significantly increased membership numbers in countries such as Malaysia, China, Singapore and Australia. The International Board operated for six years from 2002-08. During that time, the overseas membership doubled from 4,000 to 8,000.

The Board ceased to exist after the new governance arrangements were introduced in June 2008. However, by that time the traditions of the CIOB had been exported, with each country adapting them to the needs and environment of the local area. The CIOB, once London-centric, is now a truly global organisation.

Detail of Portakabins on a building site | United Kingdom

Sir Peter Trench CBE

{1918–2006}

Honorary Fellow (1983) of the Chartered Institute of Building

Peter Trench was the son of an American entrepreneur and his Scottish wife. Peter's mother died when he was a small boy, so an aunt took on most of the responsibility for his upbringing.

The young Peter elected to study economics at the London School of Economics, but his role in the Territorial Army caused him to be called up before he could graduate. In 1940, he was wounded in France and evacuated from Dunkirk by hospital ship. Back in Britain, he assisted Mountbatten with the planning of D-Day, one of the most complex logistical operations imaginable. It would prove to be useful training for his later career.

"University and the army honed his ability for strategy and analysis. There's no doubt that he could think outside of the box," says Sir Peter's son, David. "That's what took him out of the mainstream. The thing that struck me most about my father is the way he managed to become a national figure. I think that starts with being an analytical individual."

His distinguished war career, which included helping to disband the German army, earned him the military OBE. At the age of 28, he left the army as a lieutenant colonel, having also earned the Territorial Decoration (TD).

Sir Peter stood for election to parliament in 1945 as the Liberal candidate for Bradford, but came a poor third to the Labour candidate. David Trench says that his father later viewed this as an aberration, although he did remain good-humoured about the episode. In the words of his obituary in The Times, "His campaign slogan was 'Trust Trench the People's Candidate', but, as he joked, 'they didn't'."[1]

Since his earlier studies had been interrupted, Sir Peter then took up a place at St John's College Cambridge. He made friends with a man named John Glyn, who was expected to go into the family business –Bovis – despite his ambition to become a doctor. Sir Peter, on the other hand, needed a job.

"I think father was offered up in place of John," says David, "but it wasn't long before he was noticed in his own right and trained up."

Within 12 years, he had risen to become MD.

"In the boardrooms, it's always good to have someone analytical, who looks at macro level issues," says David. "That was his main contribution. I always remember him as a speaker at Guildhall dinners and conferences, championing the construction industry. I believe that his only building experience involved a petrol station where a tanker sunk up to its axles. Thankfully, his foreman got him out of trouble."

Despite Sir Peter's lack of aptitude for the practical side of the industry, his outstanding strategic vision was soon noticed.

"After the war, when the country needed rebuilding, the construction industry was a good place for a dynamic young man to be," says David. "There was so much to do, and he was at the peak of his powers at the right time."

In the mid-fifties, Sir Peter went on a four-week tour of America.

"This influenced him profoundly," says David. "He wanted to understand why craftsmen in America earned six times more than their British counterparts, but buildings didn't cost even twice as much. His analysis was, of course, that this came down to productivity. Our culture was different, but he wanted to standardise, mechanise and do something about it.

1 The Times, September 15, 2006
   www.timesonline.co.uk/tol/comment/obituaries/
   article639351.ece
   Accessed 12 January 2009

{1959}

Solomon R. Guggenheim Museum designed by Frank Lloyd Wright is completed.

{1959}

Sidney Myer Music Bowl in Melbourne, Australia is completed.

{1970}

Euston Tower in London, England is completed.

He wrote an article for the Financial Times on the subject. This was crucial to his career – it was perhaps the first thing that brought him to the eye of people beyond Bovis – because almost every conference from then on included him."

In {1959}, Sir Peter became director of the National Federation of Building Trades Employers. According to his obituary, "It was a happy appointment, coming as it did when the industry's methods and labour force, still affected by the turmoil of five years of war, were being recast to meet the demands for buildings of all kinds. Trench played an important part behind the scenes in building up the authority of the National Joint Council controlling wages and conditions in the industry. He also played a leading role in reorganising the structure of the National Federation."

According to David, "I suspect that his work on productivity also influenced the decision to invite him to sit on the prices and incomes board, which the government was using in an attempt to control wage inflation."

Sir Peter was also given a place on the National Building Agency, but this was a less happy appointment from which he would ultimately resign.

"There was a lot of correspondence," says David. "The Agency was producing reports he didn't agree with and he didn't get to see them before publication. The situation was very difficult. At the time, government was pushing for more and more direct labour, effectively making it a major contractor. My father saw this as nationalisation through the back door and a very bad idea. He just didn't see a future in direct labour."

Sir Peter remained active in a host of other areas, including ACAS (the Advisory, Conciliation and Arbitration Service) and the Post Office Arbitration Tribunal. He also continued to write and lecture on the future of his chosen industry, and served as chairman of the Building Centre from {1970} to 1973, where he once again carried out a restructuring programme.

According to Sir Peter's obituary, "He gave constant support to the Joint Consultative Committee which brought builders and the building professions together, and played a leading part in the establishment of the Economic and Planning Advisory Council which was a forerunner of the construction industry's 'Little Neddies'. In these and other ways he was constantly at work to bring builders into meaningful liaison with their professional colleagues and with government."[2]

An example of the latter came in 1975, with Sir Peter's clarion call to politicians. In difficult economic times, his address to the Institute of Building's annual conference urged MPs to consider cross-party working to deal with the problems of the industry. The speech generated significant media coverage, including the Editorial in the next edition of Building Magazine. Sir Peter argued for many things that are familiar to readers today, such as flexible skills training and constructors' involvement in the design stage. He then turned his attention to the (then radical) idea of the educated client:

"Sooner or later somebody has to face up to the fact and admit that, whereas he who pays the piper calls the tune, unless he has some slight musical appreciation, the tune can be pretty ghastly."[3]

Sir Peter, by contrast, was destined to understand the industry from a myriad of different perspectives. His multiple board memberships included the chairmanships of Y J Lovell (Holdings), James Davies (Holdings), and Building Management and Marketing Consultants, a subsidiary of the Builder Group. He was also on the board of Capital & Counties Property Company, the LEP group, Crendon Concrete and Nationwide Building Society. Knighted in 1979, he remained a regular speaker at high profile events throughout the 1980s.

Indeed, his eloquence caused one commentator to cite Sir Peter's words as an exemplar of how the parochialism of the contemporary planning regulations gave developers the moral high ground in their efforts to build. At a conference in 1981, Sir Peter argued that "Unless there is a radical change of attitude towards positive planning at local level – and this involves a reversal of the use of the planning system by the 'haves' to keep out the 'have-nots', and the need for housebuilders to rely on appeal for the implementation of statutory government policy – the prospects of providing the homes required look very gloomy indeed."[4]

A year earlier, he'd gone into print arguing that the lack of affordable homes would create difficulties for first time buyers. While this shows that his instincts for the future were generally good, he did concede his mistake at believing the eighties would prove to be the decade where off-site building really took off: "Sheepishly, I have to admit to having predicted this at the end of the last two decades and being wrong on both occasions."[5]

2 Ibid.

3 Trench, P. 1976. *Face the Future with Courage in Building Technology and Management.* January p. 21

4 Ball, M. 1983. *Housing Policy and Economic Power: The Political Economy of Owner Occupation.* London: Routledge p. 257

5 Trench, P. 1980. *Building in the 80s in Building Technology and Management.* January p. 34

{1961}

Empress State Building
in London, England is
completed.

{1984}

Neue Staatsgalerie in
Stuttgart, designed by
James Stirling opens to
the public.

From 1982, Sir Peter was also a visiting professor in building management at the University of Reading and, in {1984}, became vice-president of the National House Building Council.

According to his obituary, "Trench always maintained that the industry could be a vehicle for economic regeneration so crucial to the United Kingdom. But he argued that if it was given the economic role it deserved, it must accept change and departure from selfish sectional interest, not least from the false professionalism that divided designers from the builders."6

As one of that crucial generation who worked to put construction on a par with the other professions, Sir Peter's influence is enormously important. At a time when constructors were far lower down the pecking order than today, he showed the need for a cerebral approach – both to the business of construction and to the macro-level economic issues it reflects. He was also a great believer in education.

"Through education and persuasion, his biggest achievement was that he made the industry respected," says David.

"A lot of people heading to university would never have considered construction in those days, but they will now. That's his biggest legacy. It wasn't only down to him, but he was one of the leaders of the band. In organisations such as the National Federation of Building Trades Employers, he was mixing with the elite companies of the industry. But once you got below that, even the MDs weren't educated in those days, except in the skills. He worked to bring management techniques into the industry, and to get people to look at it as a serious place for a career, where intellectual rigour could play a big part. He believed passionately in research and development and was an early champion of the Building Research Establishment."

Sir Peter's analytical mind and seemingly effortless government connections gave him both a prominent place at almost every conference and access to the most prominent politicians of the day. But David sees the support his father received at home as crucial to his achievements.

"Mother helped brilliantly," says David. "She was very gracious and social to business friends. Many important people became good friends, and regularly came to dinner. Ministers such as Geoffrey Rippon and Michael Heseltine visited often. Mother's skills as a brilliant hostess and good conversationalist enabled that."

According to Sir Peter's obituary, "His professional and social lives were often difficult to distinguish. He was the businessman seven days a week, just as he was always witty, urbane and elegant. In {1961}, much to his embarrassment, he won the Tailor & Cutter award for Britain's best-dressed man."7

David followed in his father's footsteps and went into the construction industry, while his sister Sally built a successful career in the charitable sector. Unlike Sir Peter, however, David's contribution has been about delivering the tangible product rather than the strategy. His highly successful project management company took on the last three years of construction at the British Library before taking a pivotal role in the Millennium Dome – working to provide the 'educated client' representative advocated by his father.

As a professional in the industry, David personally appreciates the work of his father – and others of that era – to raise the stature of constructors.

"He made construction respectable," he says, "and put it on a level footing with other professions in the industry. Now, the construction manager sits at the same table as the architect and engineer. The contribution of the generation who achieved that change is immeasurable."

6   The Times, September 15, 2006
     www.timesonline.co.uk/tol/comment/obituaries/article639351.ece
     Accessed 12 January 2009

7   The Times, September 15, 2006
     www.timesonline.co.uk/tol/comment/obituaries/article639351.ece
     Accessed 12 January 2009

Sir Ian Dixon CBE

{1938–2001}

President (1989–90) of the Chartered Institute of Building

Sir Ian Dixon was the third generation of his family to go into construction, and he knew from an early age that this would be his career.

According to Sir Ian's elder son Steven, "His father, Leonard, trained as a carpenter and then progressed into site management. I think my father followed him due to aptitude rather than pressure."

Sir Ian went to the technical school in Walthamstow, and started training as an estimator. He quickly realised, however, that his talents were probably better suited to quantity surveying and changed courses accordingly.

After a brief stint with Holland Hannen and Cubitts (the commercial descendent of William Cubitt's contracting firm), Sir Ian joined Charles Foster and Sons, where his father had worked as a contracts manager.

In {1958}, the London Master Builders Federation awarded their silver medal to Sir Ian for his performance in the 'analysis of quantities for pricing' City and Guilds exam. In other words, he achieved tops marks in his year for the London region. In {1960}, he also won the National Federation of Building Trades Employers' Silver Medal for Trainee of the Year.

In {1962}, after moving to Bedford he became general manager of John Corby and Sons. He was only 23, but his ability was already enabling rapid progression.

"He was a good all-round businessman," says Steven. "He was good with people and he also had good commercial instincts."

Already active as a member in the Institute of Builders (IOB), Sir Ian knew Peter Willmott through the Eastern region meetings.

In the words of Professor John Bale, "Peter Willmott was fourth generation in a family business, but he had the insight to see they needed an injection of new ideas. The company had been around for a very long time, but once Ian joined, it very quickly became a modern company at the forefront of championing causes such as partnering and women in construction."

John also met Sir Ian through the local IOB branch and they were both on the Eastern regional council of the Institute in the late sixties.

John remembers "a very impressive, confident character. We had complementary personalities – I'm very much one to look at all sides of an issue, which is a luxury you have as an academic, but Ian had to have total certainty at any given point of time, even if he changed his mind later."

When Sir Ian joined John Willmott and Sons (Hitchin) Ltd, as the company was then known, it was a good, traditional builder with work primarily across London and the Home Counties.

Sir Ian set out to change all that, bringing new dynamism to the firm. In {1976}, he went to America to study the Advanced Management Programme at Harvard. Through the course's use of case studies, he learned that business problems were often generic rather than business-specific, or even industry-specific. This revelation would lead to a step-change in growth at Willmott's. Indeed, in recognition of his outstanding contribution to the firm's success, John Willmott became Willmott Dixon in 1987.

During one period of business expansion at the firm, Ron Malyon FCIOB, who worked at Willmott Dixon for 25 years, toured the country ensuring that any acquisitions were 'brought into the fold' and taught to work with a national set of policies and procedures.

"Ian said he wanted someone with grey hair and street cred," remembers Ron. "He selected me because I was from the grassroots, and could therefore go round and talk to everyone on a level. That's what he believed in."

Ron remembers when Prince Charles came to the opening night at the Lyceum Theatre (originally built by Builders' Society founder members Grissell and Peto), following a Willmott Dixon

{1958}

Seagram Building in New York, designed by Ludwig Mies van der Rohe is completed.

{1960}

Cairo Tower in Cairo, Egypt is completed.

{1962}

Coventry Cathedral in England, designed by Basil Spence is completed.

{1976}

Royal National Theatre in London, designed by Denys Lasdun is completed.

refurbishment. Sir Ian had met Prince Charles the week before, when he'd attended the Royal Command Performance at another venue. At the Lyceum, Sir Ian bagged the seats directly behind His Highness so he could continue the acquaintance. He also ensured that Ron sat next to him, so he could introduce the refurbishment's project manager to the heir to the throne. To Ron's surprise, Ian blithely tapped the Prince on the shoulder and picked up their previous conversation as if they were lifelong friends.

Sir Ian had charisma. As John Bale says, "he could get away with things."

This slightly larger-than-life ability to engage with and influence people was a great asset to Sir Ian's presidency of the CIOB.

"He realised the need to promote construction," says Steven, "so he launched the Building Matters campaign."

This became a fully-fledged roadshow, with a double-decker bus touring around the UK. Well-connected with the government of the day, Sir Ian prevailed on Cecil Parkinson to attend the launch event. He also contrived to get Prime Minister Margaret Thatcher photographed on the bus, surrounded by children wearing hard-hats.

This was not the only time that buses featured prominently in Sir Ian's career.

"Ian once decided that the company's board should go to site," remembers John Bale. "He had a Winnebago equipped as a board room. The experiment didn't last, but at least he tried it. It was a novel and symbolic thing to do. In fact, the symbol is more important than the reality."

This idea of bringing things closer to site was also reflected in Willmott Dixon's organisational structure, as Sir Ian removed the contract manager tier and gave more responsibility to site managers.

"He didn't believe in having a manager oversee six or eight projects," says John, "so he decided to get rid of that layer and get site managers who could take full responsibility for projects. The old pattern worked when the site manager had come up through the trades. That route created good supervisors but not necessarily managers. Ian wanted higher calibre bright young folk on site. The travelling boardroom was just one example of his philosophy of bringing talent onto site."

"Willmott Dixon was trailblazing as construction managers who were interested in the technology of their clients," says John. "Now,

there is much more interaction, but once it didn't go much further than the brief to the architect. When they started building the David Lloyd sports centres, they got so close to the client on the first project in Leeds that the same team were asked to go to build the centre in Southampton."

An early exponent of collaborative working, Sir Ian was delighted when a major project (building a factory in Paignton, Devon, for an American client – Nortel Telecoms) allowed the firm to use an early form of partnering for the construction. While he received many honours over the course of his career, including a Knighthood in 1996, the one distinction that gave Sir Ian particular pleasure was his appointment as one of the special advisors on the Latham Report. His commitment to the partnering approach was such that every one of Willmott Dixon's 1200 staff received extensive training so that it was understood across the business.

However, his focus was never confined purely to the company. As a successful local politician, he rose to become deputy leader of the Bedfordshire County Council and could have been selected for a safe parliamentary seat. John Major had great respect for him, once saying: "The thing I like about Ian is he gets things done." After giving the opportunity due consideration, Sir Ian concluded that parliament wasn't dynamic enough for him. At heart, he was always a constructor.

In fact, he had a strong sense of the direction the industry needed to take to improve. He felt that it should be more holistic, and approved of the Construction Industry Council (CIC) – where he served as chairman from 1991 to 1994 – bringing the professions together. (There is an amusing story that he once told the RICS president that "things won't get better until you bloody surveyors stop calling us bloody builders"!) .

"He believed in a sort of corporatist structure and relationship with government," says John.

"Lots of his fellow Conservatives were sceptical about this approach, but Ian believed it helped the industry to get taken more seriously. He was active in the initiatives that led to M4I, for instance. He was a liberal conservative, and a free market liberal in terms of his social attitudes. Nevertheless, he could be dictatorial if he needed to create change. He delegated, but he could certainly put his foot down. A good example would be when Willmott Dixon was sponsoring six students at Anglia Ruskin. He told his managers that they could choose any six trainees they liked – so long as three

Fortnum & Mason | London, United Kingdom

of them were women. He wasn't going to dictate the details, but he believed that if he didn't specify that women had to be chosen, it just wouldn't happen. In that sense, he was ahead of his time."

He had a strong commitment to education, reflected in his appointment as Pro-Chancellor of Luton University in 1999, as well as the CIOB scholarship that bears his name. Every year, a young professional is given the opportunity to produce a paper on a topic related to the industry and present it to an industry audience.

On his death, Willmott Dixon group chief executive Colin Enticknap FCIOB said: "It was a real privilege to work closely alongside Ian. Many aspects of his personality have become permanently engrained in our culture." But Sir Ian's influence, while significant for Willmott Dixon, extended across the industry too.

He also did a great deal for his local community – as the citation for his honorary doctorate points out. Awarded in 1992 for services to education and training, it reveals that he served as Chairman of the CBI Eastern Regional Council, the North Hertfordshire Health Authority, the Riverside Health Authority and Bedfordshire Training and Enterprise Council.

In the words of Lady Dixon,"We lived most of our working lives in and around Hitchin where Ian was involved in many aspects of local life. In 1997 he was asked to be President of 'Hitchin Town Centre Initiative' aimed at getting businesses, local authorities and community organisations working together to improve the town. We wanted any memorial to Ian to be in Hitchin. It took six years to find a suitable project but eventually the chance to replace the old Fountain behind the Parish Church was suggested, a perfect memorial."

As Ron Malyon remarks, had Sir Ian not succumbed to illness he would undoubtedly be continuing to make a major contribution to the industry. He is certainly remembered with great affection, and as a major influence, by many of those who worked for him.

"He knew how to make other people feel important," says Ron. "Willmott Dixon trainees would get invited to the company box at Ascot once a year, and he would come and talk to them all personally. It had a huge effect on their morale. He always understood that motivating people was about more than bonuses. He would send hand-written notes thanking site managers for their work on projects. He had a gift for the personal touch – for working out the best way to make each individual feel valued. That's what made him an effective leader. But the thing that impressed me most was the fact that he understood the problems which happen on the ground, on site. When you're a site worker, having someone senior that really understands that is pretty special."

Sir Ian never lost touch with what was going on at site level, and he could happily engage anyone from Prince Charles to a labourer in conversation.

"He really could turn on the charm," says Steven, "but he had no airs and graces – he never forgot what was important to people on site. He knew how to talk other people's language, and that made him a great communicator."

Sir Frank Lampl

{1926–   }

Fellow of the Chartered Institute of Building

As time passes, it's easy to forget the connotations that certain terms used to carry. But this great industry leader was once judged to be a "bourgeois undesirable" – words that are now quite anodyne, but which were disastrous for a Czechoslovakian in the mid-twentieth century.

Already a survivor of Auschwitz and Dachau, the young Frank Lampl was sentenced to imprisonment in the uranium mines at the Czech town of Jachymov. Released in the amnesty following Stalin's death in 1953, Frank was only permitted to work by labouring in either mining or construction.

Feeling that he'd already seen enough of the mining industry during his incarceration, Sir's Frank's career in construction commenced.

"In those days, people believed that a political enemy of the government couldn't expect a managerial job," explains Sir Frank. "I remember a doctor of law with a window-cleaning job. That was normal."

Accordingly, he spent the next eighteen months as a site labourer. For someone who had already spent so much of his life as a prisoner, this might have been yet another great frustration, but Sir Frank does not perceive that period as wasted time.

"That kind of experience is a great advantage, particularly in construction," he says, "because it helps you understand what motivates people. If you come through the ranks, as most construction directors did in those days, you have a much better feel for what's happening on site. Now, the world has changed. Today, you need to be a graduate; you need knowledge and the ability to take in information. But the danger is that people may not have time to understand what the workers do. I think it's important that universities do everything they can to help their students get a perspective on what happens on a construction site."

Despite the insights that working as a labourer afforded Sir Frank, he naturally wanted a role more suited to his talents. He was elevated to site foreman, but had ambitions to study and progress still further. The fact that he ultimately achieved this is testament to a level of determination and persistence that most of us would struggle to emulate.

"I had a political label which in some circles might be honoured, but in others guaranteed persecution," he says. "This made life difficult. I wanted to study and to learn, so I applied to university. I applied three times in three years, and on each occasion my application was disallowed. I had to apply for permission – everything needed permission then – which was consistently refused. But I was determined not to give up, not to let anybody get me down. It was not the easiest time, but giving up was never an option."

This seems likely to be at the heart of Sir Frank's ability to survive some of the most testing situations a person could be called upon to endure. His unwavering determination persisted throughout.

"There is a story my father told me," he says, "about a man who went round small farms collecting milk for the market. To increase his profits, he would stop by a lake on the route to the town, open his two milk churns, and water down the milk a little. One day, two frogs from the lake got caught, one in each churn. When the man got to the market and opened the churns, there was a dead frog in one. In the other, there was a live frog, sitting on top of a churn of butter. That frog had never stopped fighting."

This fable is so much at the heart of Sir Frank's ethos that when he was given his knighthood, he had a frog put on his coat of arms.

Rose Bowl | Leeds, United Kingdom

{1984}

Bank of America Center in Houston, Texas, United States is completed and opened.

{1986}

Lloyd's Building in London, designed by Richard Rogers is completed.

{1986}

Rialto Towers in Melbourne, Australia is completed.

{1986}

Glasgow Sheriff Court in Glasgow, Scotland is completed.

elements need great attention. Hopefully, during my time leading Bovis, we achieved it to some extent."

Sir Frank's principles stood him in good stead, because in {1975} he was asked to head up the company's new overseas division. He had already showed initiative during his career at Bovis, clinching the contract to refurbish Pergamon Press after seeing news of a fire at its premises and making a cold call to his compatriot Robert Maxwell. His talents as a linguist and a relationship-builder were also undoubtedly key.

The performance of the new division outperformed all expectations and Bovis International was formed, with Sir Frank as its leader. Peter Cooper, the firm's historian, gives us some idea of the scale of the successes: "In 1982 alone, 22 contracts were bagged, including a 400 bed hotel in Sri Lanka, 50 villas in the Algarve, and a £25 million pipeline in Aden. The following year, the failure of a contractor in Abu Dhabi brought in contracts worth £50 million, and contracts were also won in Yemen, Portugal, and San Francisco."

Perhaps unsurprisingly, Bovis won the Queen's Award for Exports in {1984} and again in {1986}. By this time, Sir Frank was head of Bovis Construction, and on the main board of the company's owners, P&O.

Although he grew more distant from the daily business of projects, he nevertheless enjoyed leading the firm during some landmark projects. These included the Broadgate development, which was revolutionary in its adoption of value engineering, and Canary Wharf.

However, he sees his development of the international business as the most significant aspect of his career.

"I wanted to build up a business which didn't suffer so much from the economic cycles," he explains. "My philosophy was based on the fact that economic cycles are not the same all over the world. Anglo-American countries have a different cycle from the East. If you build up a business that's more international, then you are fighting the cycle. By the time I retired, we had a presence in 40 countries."

During this time, he was also involved in winning the contract to build "Euro Disney" (now Disneyland Paris), a demanding process that forced Bovis to refine its approach and focus on the service the client required.

"The competition was huge," says Sir Frank. "So we developed much better arguments and focused on how to get close to – and satisfy – the client. It was a difficult project, but I think it strengthened the position of Bovis as a professional manager of construction."

In 1990, the year he received his knighthood, Sir Frank visited Czechoslovakia for the first time since fleeing the Russians in 1968. Visiting his old employer, the Moravian construction company, he found things had changed very little over the intervening 22 years. The organisation was the same, and so were the staff. It was very different from the rapid expansion and development he had presided over at Bovis.

Now lifelong president of Bovis Lend Lease, he is still very much involved with developments in the industry and its management.

Since he is a man who sets much store by reputation, it seems natural to ask how he would like to be remembered.

"I'd like to be remembered for fairness and for caring," answers Sir Frank. "Is that more important than the business expansion? Yes, for me it is. The expansion will remain linked to my name, I think, but that's not the most important thing for me. Executives must be capable and competent, that goes without saying, but beyond that, they should be caring and fair."

{1968}
St. John's Beacon in
Liverpool, England
is completed.

{1968}
Neue Nationalgalerie in
Berlin, Germany, designed
by Mies van der Rohe,
is opened.

{1975}
U.N. Plaza, New York,
designed by Roche-
Dinkeloo, is completed.

{1975}
First Canadian Place,
Canada's tallest skyscraper
is built in Toronto.

Like the frog that created butter, Sir Frank's persistence paid off and he was finally given permission to study. Over the next five years, he became a graduate engineer and then a site manager and a construction director.

"If you work hard and do your job well, someone will notice," he says. "Building a career is an interesting thing. I always tell ambitious young people to be careful how they treat their colleagues, to think about the relationships they're creating in the team. If you rise to become chairman, on the way up you overtake your bosses, and they become your subordinates. There are many people who don't succeed because their colleagues don't have the respect necessary to make that adjustment. If your subordinate does not like you, you won't succeed. Most success depends on colleagues, on the team. You need the right relationships, and that's too often not the case."

This emphasis on building relationships, however, was one of the things Sir Frank enjoyed most about construction. Once embarked upon his career, he never considered changing.

"It's a very exciting industry to work in," he says, "and I can't think of any other sector that could match it. Any job can be exciting if you do it well, but construction has all the elements, especially the pivotal relationships with other people. You're very dependent on the team. You can be imposed on a team as its leader, but it's far better if you grow into being a leader, by being accepted and respected. That's about so much more than just technical knowledge. There's a difference between leadership and management. A leader needs understanding, vision beyond the bottom line and a goal with ethical substance that people can be excited about and proud of."

This insight is central to Sir Frank's leadership style.

"People at the top can have large egos," he says, "but you must never say 'I': it's always 'we'. You lead by having a vision, but you win with the whole team. It has to become their vision as well if you want to succeed. So you have to like people – a misanthrope can not be a good leader."

By the age of 42, Sir Frank had survived persecution from two opposing totalitarian regimes and built a solid career in construction. That in itself is a remarkable achievement, but it's really only the beginning of the story.

The year was {1968}, and the political liberalisation of the Prague Spring was about to be brought to an abrupt halt. As the Russian tanks rolled in, Frank Lampl and his wife packed a single suitcase and left for England, where their son Tom was studying on a scholarship to Oxford University. They prepared to start over yet again.

By the early seventies, this indomitable man was a project manager with Bovis. His first assignment was Luton's shopping centre, where he demonstrated his understanding of relationships through astute subcontractor selection, creating connections that still exist more than thirty years later.

He was happy to work with small firms, he says, as long as they had a good reputation. This touches on another of the beliefs that formed the foundations of Sir Frank's career.

"I was brought up to believe that the most important thing is your reputation," he says. "If you lose it, it's hard to get back. And I believe that this is true for companies, just as it is for individuals. It's the most valuable thing that the company has. Obviously, profits are very important, but if I had a conflict between making money and damaging my reputation, I know what I would do. I would take the simple view – I could always make money the next day, but I can't repair the damage to my reputation. In my everyday behaviour, and in every negotiation, this informed my beliefs."

The overriding principle for Sir Frank was to conduct business with integrity. However, he is at pains to stress how much this principle can be good for business – as well as being the correct course of action.

"Maintaining a good reputation doesn't mean you have to be naïve in business, certainly not," he says. "Obviously, a chief executive's duties are focused around enhancing shareholder value, but the reality is more complicated than that. You have to strike a balance between the interests of clients, shareholders, employees and the community. After all, you haven't got a business without clients, or without good employees. Sustainability and ecological issues also have to be considered. It's complicated to keep a balance, but again it comes back to the fundamental issue of reputation. Caring for clients and employees, and having a reputation for doing so, is a good way of taking care of shareholders. It is so interlinked that it becomes dangerous to put shareholder value entirely on its own on a pedestal. That's the way success is measured, that is clear, but in order to achieve this success and maintain it, all these other

Canary Wharf | London, United Kingdom

Professor John Bale

{1942–   }

President (2000 – 2001) of the Chartered Institute of Building

John Bale's father, a carpenter who progressed to site manager level, was keen to impress on his son the opportunities afforded by a career in construction. As a result, John Bale became aware of the work of the Institute of Builders, as it was known then, earlier than most people.

"My father used to say that the LIOB (Licentiate of the Institute of Builders) qualification is what I needed to get, because all the bosses at his firm had it," says John.

However, as a boy, John wanted to be a journalist like his uncle, who drove a big, black, shiny Rover that seemed far more impressive than mere qualifications.

"I was persuaded by my parents to go into building," says John, but I've never regretted it. I never felt dragooned into this career – I'm not innately practical, but I have a tremendous love of the industry. "

He attended Wolverhampton Technical High School, learning craftwork, technical drawing and the history of building. He points out that the schooling he received isn't unlike the diplomas now being introduced for vocational disciplines in schools; a move he welcomes.

"Even though I was in the building stream," he says, "our class eventually produced a dentist and a cinema manager, which refutes the idea that you're closing off your options – we had a broad range of schooling."

His teacher, a man called Herbert Yeadon, provided early encouragement and inspiration, and did much to get industry placements for his students. John went to work for a local building firm in the joinery works.

"I was office assistant to the manager, which at the time was the broad equivalent to being a management trainee. I enjoyed it, but the job wasn't really leading anywhere."

However, he went to college on day release and sat next to someone who worked for British Rail. This led to his next job, as trainee surveyor in the district engineer's office.

"I did a track maintenance course alongside the gangers," he remembers. "There's always a danger that we tend to understate the sophistication of manual occupations. People who inspect and maintain rail tracks are very important people."

He had two enjoyable years there (aged 18–20), with quite a lot of responsibility, travelling all over the district doing surveys for things like track drainage schemes, and spending the rest of the time in the office dealing with drawing and specifications.

However, he wanted to learn about the commercial side of construction so he joined Alfred McAlpine as a trainee estimator and, while working there, completed his Higher National Certificate. Then, his old tutor Herbert Yeadon called to ask him to be a part time teacher. That set John on his career, although he couldn't have realised at the time where a couple of evenings a week at Wulfrun College were going to lead.

"I quickly discovered that to teach, you really have to learn something properly," he says. "Even teaching elementary materials – the basics about timber and bricks – I found I needed to do a lot of research."

John realised that a couple of years teaching could give him the in-depth knowledge he might otherwise have acquired on a full-time degree. Fully intending to return to industry at a later date, he applied for six teaching jobs, got shortlisted for three, and had interviews on three consecutive days.

{1966}

CN Tower, is completed as the first skyscraper in the city of Edmonton, and tallest building in Western Canada.

{1971}

Näsinneula tower in Tampere, Finland is completed.

{1974}

Horseferry Road Magistrates' Court in Westminster, London is completed.

{1989}

Louvre Pyramid in Paris is completed.

"On Wednesday I had an interview in Birmingham, which meant I could meet my dad for a drink that night," he remembers. "I didn't get the first job, but he encouraged me to try for the others. I got the second job I went for, in Chelmsford, which meant I didn't need to go the Guildford interview on the Friday!"

His father was pleased that his son, by then a full time assistant lecturer, was able to finish his IOB studies in {1966}, at the age of age 24. John also won a travelling scholarship from the National Federation of Building Trades Employers (now part of the Construction Federation).

"Although I was a lecturer," says John, "I always kept one foot in the industry. During one summer vacation, I worked on project planning for a local building firm. Introducing the critical path method was very new then. The only text books were American. We were only just seeing the first glimmerings of computers!"

In {1971}, Professor John Andrews suggested that John might wish to study on University College London's new Masters Degree course. John pointed out that he had no Bachelor's degree, but was told that UCL was prepared to consider people with professional qualifications, on their merits.

"Professor Andrews was way ahead of his time," says John, "because it was unusual to go straight to a second degree back then. But somehow he convinced the university senate that it would be alright, so I sat a qualifying exam and attended an interview to get in. I found the course awe-inspiring. Suddenly, I could understand construction as a strategic industry. Many people from developing countries attended the course, learning from great teachers like Professor Duccio Turin and the economist Dr Patricia Hillebrandt. Some construction Masters have tended to be 'how to do it' courses, but this was at a higher level – teaching us to try to understand the industry in terms of its organisations and its place in the economy." Immediately after completing the course, he worked with Andrews and Hillebrandt on a report for the National Economic Development Office, entitled Project management – proposals for change – {1974}, which was well ahead of its time.

John later spent eleven years teaching construction management on UCL's MSc course for half a day each week, while continuing to work at what would become Anglia Ruskin University.

Around this time, he got to know Ian Dixon, as they were on the Institute's regional council together. John was impressed by Ian's work in the transformation of Willmott Dixon and its commitment to management education; the company would later reciprocate that interest by sponsoring John's professorship for thirteen years at two universities.

By 1982, John was head of construction and surveying at Anglia Ruskin. Three years later, the then head of planning retired and a merging of the departments was mooted. John suggested 'Built Environment' as a name for the new department, which was something of a first in academia since the term was only just coming into use. The 1980s also saw John making a crucial contribution to the development of building education at a macro level.

"In the early eighties," he explains, "there were a handful of building degree courses, and not even the people running them had much confidence in expanding the discipline. The industry wasn't confident either. At Chelmsford, we ran a four year diploma, but not a degree course. Courses needed government approval and I felt that the discipline and the CIOB were being sold short. Other disciplines such as catering were expanding rapidly, but the core discipline of construction, despite the size of the industry, was not there."

He discussed his concerns with Institute secretary Dennis Neale, who asked him to write a report setting out the case, for presentation to the Board of Building Education.

"Writing the report," says John,

"I was aware there was still scepticism about degrees. The industry was worried that students would become too theoretically-orientated."

Nevertheless, John could see that it was vital to the industry for construction management to become a graduate profession. His paper, Towards a Strategy for Higher Education in Building (1984), put together some carefully crafted arguments. He showed how increasing complexity and client expectations were making higher education vital to the performance of an industry that's always been crucial to the national economy.

A source at the Ministry of Education told the delegation to "set up a committee chaired by someone important, because then we'll have to listen". The Lighthill Committee was duly formed.

"This paved the way for the expansion of degree courses," says John, "and allowed the Institute a stronger link with higher education. The timing was vital, as a massive period of growth followed. In the six years to 1992, higher education provision doubled. If we hadn't started when we did, that expansion would have left construction behind."

By {1989}, John was Dean of Built Environment, Science and Technology at Anglia Ruskin.

"The job had become more administrative," he says, "but it gave me valuable links with industry. As Dean of a wide ranging faculty, I helped to build a relationship with Ford Motor Company. When I went to see their new plant in Valencia, I ended up on Spanish TV launching their new training centre! I never wanted to lose contact with construction, but I enjoyed these other things. The construction industry can be very narrow, so I like going into something completely different such as Ford, drawing parallels, and asking: what can we learn?" Another interesting challenge was working with the radiography profession on the development of new degree courses in that discipline.

In 1995, he moved to Leeds Metropolitan University – to lead a Faculty embracing all the construction disciplines, including architecture. This then merged with Health to form a mega-faculty of Health and Environment.

"There are definite similarities between the built environment and health," says John. "Both have lots of specialisms, and both are dominated by an elite traditional profession."

Back on home turf, John chaired the Construction Industry Council's Education Committee from 1997–2000, and put substantial effort into the production of that body's education manifesto.

"It was challenging trying to get all the constituent bodies to sign up to a common education manifesto," he remembers, "and we needed commitment to joint working."

Eventually, a strategic document was agreed and produced, which was no mean diplomatic achievement. At the same time, John was working his way up at the Institute – becoming CIOB President in 2000.

As well as making a pivotal contribution to the development of construction education in the UK, John also assisted with the international standing of the discipline. He obtained EU funding to help train managers in Poland, and became external examiner to universities in Hong Kong and Malaysia. Through the CIOB, he was also heavily involved with educational development in China.

He retired from Leeds Metropolitan University in 2002 and was elected to Leeds City Council in 2004, where he now chairs the Corporate Governance and Audit Committee.

"I'm afraid I treat my meetings a bit like seminars, bouncing ideas about," he says. "In all of the work I do, I'm usually trying to understand connections between things and how human organisations work. In civic life, as in construction, it's all about ordinary people doing extraordinary things. I can't remember where that quote comes from, but it comes back to me over and over again. It's what organised activity is about. My dad used to walk around town, look up and say: 'I put that roof on'. Even in his 80's he could talk about what he did in 1938. That sense of pride and achievement is terribly important. As a teacher, I've been doubly blessed. I remember students coming for their first interview, maybe needing to be persuaded to enter construction, and I get enormous pride from what they achieve. It's vicarious, but it still gives me a buzz. This is especially the case, perhaps, because building attracts people who don't see themselves as high-fliers academically, so it's wonderful to watch them running massive projects."

Since John's former pupils include Professors Dave Langford and George Ofori, the CIOB's Michael Brown, and Peter Mack – project manager on the Scottish parliament building – he clearly has many opportunities to savour his ex-students' achievements. Even more important, however, is that John's intervention ensured that higher education didn't expand without construction. Many others contributed to this effort, of course, but he provided an impetus and vision when it was most needed.

"I don't really see myself as a researcher," he says modestly, "but I hope I've helped to intellectualise construction. That's the route whereby it will get the recognition it deserves."

The Petronas Towers | Kuala Lumpur, Malaysia

Alan Crane CBE

{1945– }

Fellow of the Chartered Institute of Building BMYA winner 1981

Alan Crane has a pet hate. He doesn't like people saying something 'can't' be done. Whether he's arguing for tougher safety measures in the construction industry or getting projects completed in adverse weather conditions, he always likes to look for the positives.

"It's about what you can do," he says. "I don't spend five seconds on what I can't do. I don't even stay in the same room as people who tell me what they can't do. We don't have enough time on this earth as it is. We have to take the time we've got, and spend it on positives, spend it on solutions. "

This, he says, rather than ambition per se, is what's fuelled his meteoric rise through the industry.

"I've never sought promotion," he says. "I just see what needs to be done, and then I go and do it. I've always worked on that basis."

After leaving school, the young Alan Crane spent a couple of uneventful years in diabetes research. By chance, when he was between jobs, he was offered temporary work as an industrial painter.

"I've painted everything from pylons to ships and storage tanks," he says. "To be honest, a lot of it was pretty horrible. The work could be dirty, cold and dangerous. Safety standards didn't really exist in those days. Although I wasn't trained, I often had to rig up my own bosun's chair – at 200 feet up! I remember one heart-breaking, back-wrenching job where I had to use an 18-inch paint roller to put red lead on the outside of tanks. That was highly dangerous. Even then, in the sixties, lead was known to be a health hazard. I was given gloves and a pint of milk a day, because some nutcase came up with the theory that milk got

lead out of your system. Eventually, I said to the boss: 'For God's sake, there must be a better way. Surely there must be a way to spray this on.' My boss said I could talk to the paint sales rep. Then, I found coatings could indeed be sprayed on, and that this could do the work of ten men."

In his younger years, Alan had strong socialist leanings, but he was very in favour of this innovation – despite the reduction in the workforce it implied – because of the safety benefits.

"There was plenty of work anyway," he says. "Before the 1980s, there wasn't the same emphasis on health and safety, but it's an issue I've always banged on about. I'm very passionate about ensuring that we think of health issues in the widest sense. In the sixties, I used to spray asbestos coatings using just a paint mask for protection. Safety is what people talk about, but health related issues like asbestos kill more people than accidents. There are more people who will never work through health than safety."

Much later, Alan would fall out with a colleague at the Construction Federation over his views.

"A few years ago, a target to reduce fatal accidents by 10% was announced," he remembers. "I took issue with that. How can we possibly have a target that accepts that we, as an industry, are going to kill 70 people next year?"

He says he's never forgotten the situations he encountered as a young man – particularly the dizzy heights scaled in his bosun's chair. But despite the difficulties and dangers of Alan's early years in construction, his ability to deliver results was noticed. He was put in charge of small jobs to begin with, then larger contracts were entrusted to his supervision and he was trained in estimating and surveying. On promotion to foreman, he started to study in his spare time. He was made a supervisor, and then an area manager. By the early seventies, he was regional manager for Cameron Industrial Services Group.

"I focused on three or four key customers, and worked on building better relationships," he says. "I found it was more efficient that way. Customers would start talking about a range of work they

{1981}

JP Morgan Chase Tower, Houston, Texas designed by I. M. Pei is completed.

{1981}

Sydney Tower in Sydney, Australia is completed and opened.

{1981}

Colonius in Cologne, Germany is completed.

{1981}

The Riyadh TV Tower in Riyadh, Saudi Arabia is completed.

needed completing. I found I was providing everything from joiners to suspended ceilings."

One of the customers Alan cultivated was Bovis. They were so impressed by his attitude that they invited him to join them as a project manager. When Alan said he wasn't qualified for the role, they pointed out that he was a project manager already in all but name.

"I think it's the one time in my life I said I couldn't do something," says Alan. "And they just told me not to be so daft!"

This was the turning point in his career. His income fell sharply overnight, because his old job gave him performance related pay, but he could see that in the longer term he'd do better at Bovis. This indeed proved to be the case.

By {1981}, he'd won the CIOB's Building Manager of the Year Award, which led to Alan's continuing involvement with the Institute.

"Then, there was only one category," he says. "Instead of a big awards ceremony, there was a lunch at Quaglino's in London. I was proud, over the moon, in fact, so I pitched up for the lunch."

There was then a discreet conversation between Alan and a stalwart of the Institute, which eventually led to the membership qualifications being changed completely.

"Someone came over and very quietly whispered 'You're not a member?' remembers Alan. "I said 'No – does it matter?'"

Similar conversations had probably already taken place, because it had been deemed within the rules for Alan to receive the reward. However, when Alan was asked why he'd chosen not to join, he explained that since he'd come up 'through the tools', he didn't now have the tools required! He also pointed out that 50% of the people running construction projects wouldn't qualify either.

Later, this conversation led to Alan being invited to join the Professional Practice Committee – where he is now Chair and instrumental to developing more routes into membership.

"I wanted to create a way for anyone in the industry who wants to join," he says.

"Everybody working in the industry is a professional. I'm quite proud about my contribution to widening the route to qualifications. Before, most people couldn't get in."

Alan himself duly gained full CIOB membership via the new DMX exams, which assessed his experience rather than prior qualifications. He also won the Queen Elizabeth II Silver Medal for his work as project director at Gateway II in Basingstoke.

According to Alan's citation, "he quickly acquired a grasp of the design concept and he and his team were able to make constructive suggestions while an innovative design was still not final… Problems, including new and unusual techniques, were overcome in an expert and confident manner and the building was constructed fast and well."[1]

"That was a very special job," says Alan. "The client employed a really switched on architect from Arup Associates, a fully integrated design practice. The client gave us permission – almost demanded us, in fact – to try to do things differently. It was the first building in the UK with fully panelised, prefabricated curtain walling. That's the norm now, but it wasn't then."

The element of prefabrication was helpful, as his team were working through the worst winter in living memory.

"At one point, it was the only project in the south east of England still progressing," he recalls.

Christmas 1981 came and went without the traditional break, and only three days were lost due to bad weather. Normal contractual provisions for inclement conditions were replaced with technologies such as portable shelters, but it was Alan Crane who kept the team motivated in freezing conditions. The project was also noted for its successful use of natural ventilation, and such groundbreaking construction technologies as the micro-computer and the cordless telephone.

By this time, Alan was Divisional Director at Bovis and would progress still further, eventually becoming Chief Operating Officer for International Building. In this role, he was responsible for major projects including EuroDisney, Kuala Lumpur City Centre (The Petronas Towers), part of the Barcelona Olympics and a variety of BOT/BOOT schemes in Malaysia, Indonesia and China.

"I was living on aeroplanes. It was fabulous," he says. "What a privilege for a cigar-smoking kid from Dorset."

One of the other major projects he directed during his rise through the ranks was the Canary Wharf development.

"That wasn't just important to me, it was important to the industry," he says. "It caused the economic regeneration of a major part of London. But the project was a huge challenge. The location made sense when you looked at the tax breaks, but not from any other perspective! There was no land. We literally had to create the land over the water, creating buildings three times the width of the existing wharf. We put piles down into the water and built out over it, to create the basis for the development. We pumped out the water and built up from the floor of the dock. We put a railway in as well as a road system. Power, water, sewerage – there was nothing there."

This achievement is a source of great pride to Alan.

"The project shows why I love the industry," he says. "Every day, you go out and do something different. Some people can't cope with not knowing what they'll have to do. I love it. In construction, you have to think on your feet all the time. The industry's given me constant opportunity to do exciting things. I think I'm so lucky, because I don't have to do boring things I don't want to do. I can do exciting things all the time!"

He regularly goes into schools to give talks to young people, stressing that there's no limit to what can be achieved within the industry.

"There's so much opportunity for advancement," he says. "And there are very few industries with such job satisfaction. You leave a lasting legacy. I'm one of the luckiest men in the world. There are dozens of industries where you can make more money and have a more stable existence, but what does a financial services advisor leave behind?"

"All that innovation means is doing it differently," he says. People confuse innovation with invention, and it's not the same thing."

That's not the only involvement Alan has had with leading industry bodies. As befits a man without particular career ambitions, he didn't set out to become Chairman of the Construction Confederation.

"The Confederation started because the government was fed up dealing with umpteen organisations, and wanted one umbrella organisation," he explains.

It was also the time of the Egan report "Rethinking Construction", and the government wanted the industry to provide better value for money. He thinks that his name came up because the time was right for someone who would question the Industry Establishment and existing ways of doing things.

Given his enjoyment of finding new ways of doing things, it's perhaps natural that he ended up being asked to lead M4i, the Movement for Innovation, which was tasked with getting industry to adopt the recommendations.

And it was for innovation and services to the construction industry that he received his CBE from her Majesty the Queen.

"I think some people were unhappy because they thought I'd force them out of their comfort zone," says Alan. "Personally, I can't remember ever having a comfort zone. I'm not comfortable with anything. Everything we do, we can do better. We've still got a long way to go. We're getting better, but we should deliver really good quality to all customers, all of the time. People who don't care wind me up. That's been my subconscious theme all my working life – right back to the bosun's chair!"

Nothing, however, dims Alan's pleasure at the palpable legacy of his working life.

"When I'm out driving with my kids and grand-daughters, I can point out buildings I've worked on all over the country, all over the world in fact," he says. "I've got a legacy in 80 countries. It's a privilege, but it's also a great motivation for doing things properly."

New York skyline | Unites States of America

Professor Xu Ronglie

{1931–   }

Honorary Fellow (2001) of the Chartered Institute of Building

Professor Xu Ronglie was born in May {1931}, to a father who was a businessman and a mother who was in farming.

Although there was no tradition of construction in the family, Professor Xu remembers that as a small boy he was fascinated by the traditional tales of China, especially those about the great construction projects. At first, it was stories and pictures of the Great Wall, built thousands of years ago to protect the nation from her enemies, which captured his imagination. The achievements of the past gave the young Xu Ronglie a pride in Chinese construction that still abides to this day.

He also recalls learning the widely-told legend of Da Yu, who led locals in the Yellow River Valley to build a water defence system some 4,100 years ago.

Da Yu was put in charge of stemming the Great Flood of China. Since his father had been executed for failing at the same task, he had every incentive to succeed. Whereas his father built dykes that collapsed time and again, Da Yu developed an ingenious system of channelling the water which survives until today despite being close to the epicentre of the 2008 Sichuan earthquakes. Not only did he save a population from disaster, he became the first ruler of the ancient Xia dynasty.

As a child, Xu Ronglie felt very proud of Da Yu, who came from the same part of China. Today, he says that "the legend reflects the wisdom, capability and perseverance of the ancient Chinese." Ingenious and successful leadership, reflected in an innovative and successful engineering project, was clearly part of his frame of reference from a very young age.

Professor Xu also displayed an early aptitude for maths, earning a place at the prestigious Shaoxing City Middle School of Zhejiang Province. His maths teacher, another important early influence on his development, appointed him class leader. He was greatly encouraged by this vote of confidence and started to take an interest in learning 'leadership'.

Top scores in maths would also mark him out during his university studies, and he was elected Class Leader and, later, President of the Student Union.

"Maths helped me in logical thinking," he says, "but the experience in the Student Union enhanced my capabilities in communicating with people. Communication is the skill I regard as the most important in becoming a leader."

Despite his exceptional talents in pure maths, Xu Ronglie had not elected to go to university to study the discipline. His course was in Civil Engineering. The past had inspired him as a boy, but as he got older he was enthralled by modern engineering achievements.

"The Shanghai International Hotel had 24 floors and was the highest building in China at that time," he says. "I was impressed by this, and even more amazed to see pictures of skyscrapers in USA. By that stage, I was totally attracted by construction and wished to become a builder of the future. Studying Civil Engineering was therefore the natural choice."

He still vividly remembers his industry experience in the second year of university. The project was to pull down the shanties where the poor lived in Shanghai city and then build a residential area called 'Kong Jiang New Village'. In Kong Jiang, the shanties were replaced with buildings of concrete and brick that were three to four floors high.

"The project was huge," he says, "with around 100,000 families relocated. I was excited and proud to witness the great benefits to those families, all due to renovation and construction"

After graduating from the Department of Civil Engineering at China State Nanjing Institute of Technology in 1953, he went on to study at the Academy of Architecture and Construction in the Soviet Union. From 1956 to 1958, he worked as a researcher in the Department of Foundation and Underground Structure of the Soviet Union Building Science Research Institute, studying deep foundations and underground structures. After returning to China, he then gathered work experience on site before moving to government.

Although highly gifted in technical fields, he also found that his working life constantly demonstrated the importance of good management and leadership. His motto is 'diligence plus wisdom', which he regards as essential qualities for a good leader.

Asked about his own leadership qualities, he says he's good at grasping opportunities, but also focuses on approachability, tolerance and thoughtfulness.

He quotes Confucius: 'Do not do to others what you do not want others to do to you', a maxim which he has always tried to apply to his working life.

"Respect for people wins you respect and support from others," he says, "which has to be demonstrated through effective communication."

{1982}

Georgia Pacific Tower in Atlanta, Georgia, United States is completed.

{1987}

One Liberty Place in Philadelphia, United States is completed.

{1987}

The tower at Stade Olympique et La Tour de Montréal in Montreal, Quebec, Canada was officially completed.

{1990}

Bank of China Tower in Hong Kong, designed by I. M. Pei, is completed.

Confucianism has informed the structure of society in China since 200 BC. Political and social changes notwithstanding, it still leaves its mark on the culture – as a concern with collective interests and virtue rather than individual achievements. Therefore, even though Xu Ronglie's technical abilities are indisputable, it is unlikely that he would be viewed as an outstanding leader without his innate consideration of others. As our translator, Liu Mengjiao, points out, perfectly executed projects in China are recognised with plaudits for the team, rather than for the leader. This means that the hallmarks of success in China equate to a more modest paradigm of leading by virtuous example, rather than the Western cult of the personality.

While all careers are, in some sense, shaped by what's happening in the world, we often only notice this at times of recession or crisis. But for a rising star in the tumultuous political climate of the mid-twentieth century, it would have been impossible not to be fully aware of the part one's work was playing in global events. Like many constructors and engineers of that time (and throughout the centuries), Xu Ronglie found that his country needed his expertise to help strengthen their military capabilities. As a young man, this would test his leadership skills to the utmost.

He was seconded for six months to a secret project, one that was so confidential that not even his family could know where he was or what he was doing. The only civilian member of the team, he participated in the construction of the country's first continental missile launch station, located in Jiu Quan Aviation City in Gansu Province, northwest China. Responsible for technical issues such as groove digging and sedimentation, he had to work with the team under severe weather conditions, in temperatures sometimes as low as minus 40 centigrade. Nevertheless, despite all the difficulties, the project was a success – and he remembers being full of excitement and pride when China's first continental missile went into space in 1960.

In 1968, he was appointed Chief Engineer on the project to build China's first nuclear ship factory, which would test his practical working capabilities and management skills to the limit once again. In the process of building the biggest workshop on site, the team had to hang and install steel frames – each with a 33-metre span and weighing 20 to 30 tons. At the busiest time, there were over 2,000 people and eight cranes working on site, assembling, hanging and then installing the frames.

"It was a big challenge, not only technically, but also in managing the great number of workers on site," he says. Despite having been involved in far more research than site work before that time, the project was a success and marked a turning point in his career.

By {1982}, he was the first Chief Engineer of the Chinese Ministry of Construction. He found that a new set of skills, in the analysis and integration of information, were key to his success in the post.

As he says, "it's important to look at both 'hard' information on technical issues and 'soft' information – the human dimension, which used to be neglected. I spent a lot of time collecting information about people's skills and professional experience and used this to organise projects by appointing and allocating the 'right' sort of expertise."

He also led the team that created China's Construction Technical Policy, popularly known as the Blue Book.

"The book set technical policies on development trends at a macro level and was the first of its kind," he says. "It was awarded first prize for Science and Technology Advancement in the Ministry for {1987}."

His research output led him to develop numerous policies, from minimum sizes for residential buildings to a variety of environmental issues. Thanks to his work, China had policies on waste landfill, sewage disposal and the recycling of materials before {1990}. In recognition of his achievements, he was elected as a Panel Member of China's National Science and Technology Assessment Committee and Vice Director of the Drafting Committee for China's Seismic Intensity Scale.

Professor Xu was Chairman of China's Society of Civil Engineering (CSCE) in the 1990s and was proud to establish the nationally renowned 'Zhan Tianyou Award in Civil Engineering' in 1993. Zhan Tianyou was one of the greatest characters in recent Chinese history, respected for inaugurating the country's railway construction business, for pushing forward the development of science and technology, and for becoming the first Chairman of CSCE. Given to the best creative and innovative construction projects nationally, the Zhan Tianyou has become one of the sixteen national-level key awards recognised and approved by the State Council of China.

Professor Xu would win another prestigious award himself for his English-language book, 'The Theory And Practice Of Piling And Deep Foundation In China', an edited collection of Chinese expertise in the topic.

Following forty years of research, practice and leadership, Professor Xu retired in 1994. However, for many lifelong achievers, retirement does not equate to a cessation of professional activities, and Professor Xu is no exception. He served a three-year term as the first Chairman of CIOB China, and strongly believes in the role of education in professional practice.

"Good basic knowledge is essential for problem-solving – on site, you seldom meet the same problem twice," he says. "You need basic study, and then you need to learn from experience. It's also important that, as an industry, we continue to learn from technical advances in related disciplines."

Even now, in his late seventies, Professor Xu is engaged in research projects into both management and technical problems. He also undertakes consultancy roles in projects and companies, providing advice in management (e.g. the renovation project in Taipei city) as well as in technical fields (e.g. the sea reclamation project in Zhejiang province). He is also Consultant of The Chinese University of Hong Kong, a part-time professor at Tsinghua University and Tongji University and MSc tutor in Xi'an University of Architecture and Technology. He served as president of the Chinese Association for Construction Mechanics, deputy president of the Architectural Society of China, and consultant to the Chinese Civil Engineering Society. His work has led to honours from institutions around the world, including Princeton University and the Royal Swedish Academy of Sciences.

How does he find the energy? Well, as our translator points out, Professor Xu looks the picture of health and maintains a disciplined lifestyle. Since his retirement, he has gone swimming every afternoon, and plans to remain healthy and active in the industry for years to come.

"Retirement hasn't ended my concern for the development and progress of the construction industry," he says. "I still want to help my friends in the field."

Although he stresses the need to understand construction's long tradition for slow and organic development, he is clear that the current climate necessitates a step-change in the way we embrace technology.

"In China, we have to meet the challenges of a fast-developing economy, and produce innovative mega-projects efficiently," he says. "The development of new materials, new structures and new techniques is crucial. In a system undergoing rapid change, we need to develop fresh ways of working. Innovation has to be our focus. At the same time, more attention should be paid to sustainable development. It's a major challenge, but we have to be more careful with our use of energy and materials to protect our environment for future generations."

Dr the Honourable Raymond Ho Chung-tai MBE

{1939–   }

Honorary Fellow (2004) of the Chartered Institute of Building

Raymond Ho became a leader early in life. Studying at Hong Kong's largest and (very well-known) Chinese school, which had eight classes (or 400 students) per year, he ranked first in achievement in his whole year group.

Academically distinguished from then onwards, he became class representative and then prefect. Apart from class prefect, he was appointed as the founding chairman of the permanent society for that particular grade and the society for all students graduating in the same year at high school. Not only is it clear that responsibility came preciously young, but also – according to Raymond – that he's been busy ever since!

As a young man, he developed a nonchalant attitude that belied the hours of study he put in. This created an impression of effortless achievement, exaggerated by his pursuit of a range of extracurricular activities. Interested in music, culture and sports from an early age, he became chairman of a number of school societies and interest groups, and trained to become the conductor of the school's harmonica orchestra.

"There could be as many as thirty musicians with different harmonica instruments," he remembers, "but I trained my ears to be sharp enough to lead the orchestra. There were many subjects that caught my interest, including astronomy, philosophy and psychology. I also liked drama. But my father said that studying these subjects would not bring me a good living. He encouraged me to be a civil engineer, saying that there is always construction, demolition and more construction, or alternatively, a medical doctor."

Raymond's exam results were so good that he received a scholarship after the Public School Certificate Examinations and could have studied any subject at university, but he followed his father's advice.

While native intelligence and application were crucial components in his outstanding academic record, Raymond also has an exceptional memory, particularly for numbers ("I remember phone numbers for quite a while if I've dialled them once," he says). This made him naturally interested in maths, but he retained his voracious appetite for knowledge in other subjects beyond his school years.

"I took a lot of interest in whatever I could lay my hands on," he says, "and always joined organisations which were not related to engineering when possible. For example, I joined the Red Cross, and served on the committee looking after schools for physically-handicapped students for five years; I also served the St. John's Ambulance, working all the way up to its Council, for 20 years. I joined the Friends of the Scouts and the Rotary Club, becoming President of my club in {1985}. They have since honoured me as an Honorary Rotarian."

Engineering organisations featured as well, of course, and as an undergraduate at the University of Hong Kong he became vice chairman of the University's Engineering Society.

In 1963, as soon as he'd completed his first degree, he moved to the UK to study for postgraduate diploma in Soil Mechanics and Foundation Engineering at the University of Manchester. After graduation, he spent four years with a major engineering contracting firm, Taylor Woodrow Group, in the UK and obtained his professional qualifications in the meantime. Six months after he joined the company, Raymond was already being put in charge of projects with a high degree

{1973}

Sydney Opera House, designed by Jørn Utzon is opened.

{1973}

Tour Montparnasse in Paris, France is opened.

{1973}

Cromwell Tower in London, United Kingdom is completed.

{1985}

Columbia Center (formerly the Bank of America Tower) in Seattle, Washington, United States is completed.

of complexity or with a near-impossible contract completion time – in other words, the projects that others tried to stay away from!

"I tended to be asked to do the jobs which were rushed or complex," he says. "At the Barbican in London, for instance, I was looking after the construction of a building with many complex aspects and technical constraints due to the adjacent underground station and the tunnel connected to it."

This brought its own rewards, however – he remembers getting three pay rises every year without asking, which meant his salary far outstripped that of his contemporaries.

Despite this, London's City University approached him with such an attractive research grant offer that he could not refuse. He decided to move back into academia in 1968, plunging himself into a research area which the previous researchers failed to come up with anything on. Within an impressive two and a half years, he had successfully completed a PhD thesis in "Bond Performance and Failure Mechanism of Tensile Lapped Joints of Deformed Bars in Concrete Beams". The university asked him to stay on to become a lecturer, saying that he could become a professor before he was 35, but Raymond declined because he wanted to go back to the industry.

During this time, he also organised a society, with himself as the founding chairman, for Hong Kong professionals working, doing research or studying in the UK. It had over 100 members, who would come from all over the UK to London to meet other Hong Kong people at the Society's functions.

"Wherever I go, I always start something!" he says.

After finishing his PhD, he joined a consulting firm as an Associate Partner with free share-holdings. Although he was soon invited to join the Partnership, he declined because he wanted to go back to Hong Kong at the end of {1973}.

Upon returning to Hong Kong, he joined a new-comer, Maunsell. Together with three expatriates, he built up the firm from about 20 staff to 1,000 in only a few years. Having been a Senior Director for 18 years, he found that he had already done all types of engineering projects, including all the major infrastructure works of two new towns (Shatin to accommodate a population of 600,000 and Tseung Kwan O to accommodate 400,000, with over 100 contracts), all the new KCR railways stations from Kowloon Tong to Lo Wu (the only railway line at the time and now called the East Rail) and the associated bridges and roadworks. He also had direct involvement with projects of tunnels, bridges, flyovers, roads, wharfs, hospitals, hotels, incinerators, high-rise commercial and residential buildings, geotechnical work, environmental work and also project management.

"In Hong Kong, large scale major infrastructure projects can be completed in a short timeframe," he says. "The complexity and magnitude make these projects very challenging. One year's experience in Hong Kong is better than two or three years in other countries."

He enjoys the memory of a time on an earlier project where the depth of the piling allowed him to descend 33 metres along a fault line! "It was quite an experience," he laughs, "I can always say I've done that, although I would have liked a photo. But at that time, we didn't know how to take pictures in such a deep hole in the ground!"

At that time, he suddenly found he would need to look for new challenges again. He first of all left Maunsell and spent eight months with a publicly-listed contracting firm, UDL, as Joint Managing Director. He was responsible for all engineering contracts in the Group, including projects related to the building of Hong Kong International Airport, such as the drainage and reclamation of the land for Chek Lap Kok and the assembly and painting of the 50,000 tonne structural steelwork for the deck of the Tsing Ma Bridge, leading to the Hong Kong International Airport.

"It has a main span of 1,377 metres, and is therefore the world's longest over-water bridge to carry both vehicular traffic and railways," he says. "Steel units of 1000 tons each were fabricated, barged to the site and then hoisted up to the bridge deck level and assembled together, followed with sophisticated painting work."

Cityscape | Hong Kong, China

They used an extremely long cable to span between the two towers of the bridge, with a catwalk underneath, which was removed immediately after the completion of the bridge. Raymond walked the catwalk from tower to tower.

"The highest point was 200 metres above the water," says Raymond. "I enjoyed looking down to the depths through the net! Naturally, I did this responsibly, and used a harness at all times in order to be safe."

Raymond's career in engineering clearly fulfilled both a need for achievement and for adventure. And yet he still felt the urge to pursue new outlets for his talents.

"I have been a contractor and a consultant in both the UK and Hong Kong," he says, "and even did research at a university in the UK. Then, I wanted to be a legislator! That was 13 years ago."

Now, Raymond is a member of Hong Kong's Legislative Council, representing the Engineering Functional Constituency.1

"I call myself an engineer-cum-politician," he says. "I'm often scrutinising bills, whether or not they are related to engineering, from the Rail Merger Bill for the two railway corporations to the Interception of Communications and Surveillance Bill. Being a legislator, there are two main areas of work. You have to monitor government performance and establish laws. The fact that I am using my professional experience and knowledge is the thing that enables me to make an impact. For many years, I have been chairman of the Public Works Subcommittee, which approves all government infrastructure projects. These are worth tens of billions of HK dollar every year. I recall that we approved 71.1 billion HK dollars worth of funding in a year, and 100 projects in another."

In the wake of the recent worldwide economic crisis, Hong Kong's ambitious infrastructure programme is likely to be accelerated in order to create employment and ensure stability, making Raymond's expertise more important than ever.

---

1 Hong Kong does not elect all of its representatives from geographical constituencies. Instead, some take responsibility for a specific area of interest, otherwise known as Functional Constituencies.

Raymond is also Vice Chairman of the Panel on Constitutional Affairs and now also Chairman of the 22-member committee investigating Lehman Brothers-related minibonds and structured financial products.

Raymond is also one of 36 Deputies representing Hong Kong to the Tenth and Eleventh National People's Congress (NPC) of the People's Republic of China (i.e. the PRC's Parliament). Sometimes he attends the Standing Committee of the NPC.

"We monitor the performance of the government and scrutinise bills," he says. "In many ways it's similar to my work in Hong Kong – except that the chamber is enormous, housing 3,000 Deputies, and is more disciplined!"

He is taking particular interest in areas like energy conservation, cross-border infrastructure and the waste recycling industry, as well as assisting Hong Kong people who get into trouble in legal cases in the mainland. Some of those cases can be very complex. At the same time, he is pursuing other causes close to his heart – the training of young engineers and the reconstruction of Sichuan Province in China Mainland after the 8.0-Richter scale earthquake which took place on the 12th of May 2008, claiming 90,000 lives.

"At the moment, it's one of my main concerns," he says. "I am worried about young people today – they have to try to plan for the future in a world that's changing so fast. They could lose direction. One of the ways I've tried to help is by organising many young engineers to be involved in the post-earthquake reconstruction programme in Sichuan. This is a way to inject into them the very important community spirit. Helping others brings a sense of perspective – and dealing with the aftermath of such catastrophe teaches us all that time is valuable, and not to be wasted."

Raymond certainly isn't wasting time. "I'm more than fully occupied," he says. "I've just been re-elected for the fourth consecutive term as the representative of the Engineering Functional Constituency in the Legislative Council. I stand for re-election every four years. I'm the only one among all the 30 functional constituencies to have to face fierce competition every time and still remain unbeaten since the changeover of sovereignty. The fact that I keep winning every election since {1996} gives me great satisfaction!"

Through the years, Raymond has done plenty of public services in the engineering profession, tertiary education, industrial development, science and technological development. The City University of Hong Kong offered him an Honourary Doctorate of Business Administration, the University of Manchester in the UK offered him an Honourary Doctorate of Laws and the University of Hong Kong offered him an Honourary University Fellowship, in recognition of his many achievements. Yet there are still a number of ambitions left for Raymond to fulfil.

"There are so many books I want to write," he says. "Maybe when I'm in a wheelchair, I hope I'll finally get to study my first love, which is philosophy and psychology. I want to do a cross-cultural study, looking at Confucius and Mencius and comparing their thoughts with the work of Rousseau, Aristotle, Plato and Socrates. I also want to learn more foreign languages."

This might seem a major departure from engineering and politics, but to Raymond, all things are related.

"When I studied engineering," he says, "I found that the principles are applicable to all aspects of life. Conversely, I've found that philosophy and psychology are indispensible to the practicing engineer. There are always opportunities to study and learn. I always say to people, whether you're talking to a young boy or an old man, there are always things to learn from them.

When I talk to people, I try not to have preconceptions. I'm always open-minded. I'm an open gate, all the time. If it's kept closed, nothing can come in. That's one of the most important lessons we can learn."

Sir Michael Latham
{1942– }
Honorary Fellow (1995) of the Chartered Institute of Building

A career in construction can be formed by accident as well as by design, but few of the industry's leaders are likely to have seen chance play as great a role in their professional lives as Sir Michael Latham.

He read history at Cambridge in the early sixties before moving to Oxford for a Diploma in Education. If things had transpired slightly differently, he could easily have become a noted educational reformer. Instead, fate intervened when he was offered a researcher's job by the Conservative Party. The salary was attractive, so in {1965} he took the post that set him on a completely different trajectory.

He was just completing his first week in the role, mainly doing odd jobs in the library, when Edward Heath was appointed party leader. This led to a major shake-up in the department where the young Michael Latham worked. Summoned to see the director, Michael was informed that he'd be taking over the brief for housing and construction on the following Monday, even though he admitted to knowing nothing whatsoever about the area.

"Well, you've got the weekend to learn," was the encouraging response.

To this day, Sir Michael says he never worked as hard as he did in those early weeks. He went home with books and files every night, working his way through a mountain of information on his new area of responsibility.

He proved a quick learner, and being entrusted with important and urgent tasks quickly became a marked feature of his career. When the section on housing in the Conservatives' 1966 manifesto was deemed unsuitable, Sir Michael was chosen to redraft it – in just two hours! He also remembers being asked to produce a 70-page briefing on the constitutional position of Northern Ireland in a mere three weeks. Drawing on his historical knowledge, his report gained the approval of the attorney general despite the incredible deadline.

In {1967}, he moved to a new role and became the parliamentary liaison officer at the National Federation of Building Trades Employers. He was there for six years, although in {1971} he also became Director of the House Builders' Federation. Much of his time was taken up with lobbying, trying to get successive governments to understand the difference between contractors and house builders.

"Even today," he says, "few people really understand the difference between the sectors. House builders are primarily interested in the planning process. Later, when I became a director at the house builder Lovell Homes, I don't actually remember ever discussing building costs. We focused on finance, planning, land, marketing and sales. House builders are essentially estate developers. The actual building work is often sub-contracted. They're perceived to be part of the construction industry, but they think and act very differently from contractors. They've bought the land, they're working for themselves – so they've got a wholly different attitude."

By the time Sir Michael became a director of Lovell Homes, he had become the Member of Parliament for Melton. Although his directorship would keep him in touch with the industry, as did his role as vice-chair for the parliamentary housing committee, the following eighteen years were largely taken up with work for his constituency.

"MPs have umpteen things to do every day," he says. "You deal with constituency issues from pensions to farming, you're trying to get

{1965}
NASA Vehicle Assembly Building in Cape Canaveral, Florida, United States is completed.

{1965}
Elephant House at London Zoo, designed by Hugh Casson and Neville Conder is completed.

{1967}
Habitat 67 in Montreal, Canada designed by Moshe Safdie as part of Expo 67.

{1971}
Department of the Environment Building, designed by Eric Bedford, in Westminster, London, was completed.

questions asked in the House – there's no way you can concentrate on one thing. Ministers can, but MPs can't. That's the practical reality of life in the House of Commons."

Dedicated to serving his electorate, he was always sure that he did not want a ministerial career. When it became clear during the course of 1979 that the Conservatives were likely to win the coming election, he told the Chief Whip that he did not want to be in the new government, since he saw his first duty as being to his young family. Later, he was told his decision had not been well received by the new Prime Minister, Margaret Thatcher. Sir Michael merely pointed out that out of the 350 Tory MPs, 349 wanted to be in the government, so they could do without him. He remained on the back benches throughout his parliamentary career.

His most notable work, however, was to come later. Almost by chance – not for the first time in his career – he found himself charged with an urgent and important task, which would become widely known as the Latham Report.

In fact, it was anything but a foregone conclusion that he would be the author of this influential study. Sir Michael himself was not expecting the job. However, he clearly remembers the build-up to the report being commissioned.

"In late 1991," he explains, "the Building Employers Federation held their biannual conference. They called it Building Without Conflict, but all the speakers from the different stakeholders in the industry were blaming each other! Nothing happened immediately, but early in 1992 the forerunner to the Major Contractors Group held its AGM. This was significant, because Neville Sims, CEO of Tarmac, spoke at the meeting and called for an enquiry into the problems of the industry. Sir George Young, the Minister for Construction, was there as a guest speaker, and he agreed to the proposal. Eight months later, following protracted discussions with many industry representatives, the terms of reference were established. I remember that these took up a whole page – with basically a sentence for every interest group consulted!"

It was agreed that the report would have a single author rather than a committee – some groups believed it would then be easier to distance themselves from disagreeable findings, should the need arise – and the hunt began for someone to write the report. According to Sir Michael, many retired secretaries of state and similar luminaries were invited to be the author, but all declined.

"I was the only person prepared to do it!" he says.

He was paid two days a week to write it, but instructed to give up his other commitments for the duration of the project. At the time of writing the report, in the early nineties, the industry was in severe recession. He remembers people joking that things were so bad, they would try anything he suggested.

As the good times returned, however, many people claimed that his key recommendations – for collaborative working and partnering – were not needed. There was money to be made on old-fashioned tenders.

"That might be true," says Sir Michael, "but if you don't give the client the best work possible when things are good, they won't come back in the bad times. We have to take the long view."

As an historian by training, the long view comes naturally to Sir Michael. He was acutely aware that he was producing recommendations for an industry that had changed profoundly over the previous generation: the discipline of construction management had developed and specialist subcontractors had become key in the way things were organised.

Called 'Constructing the Team' and published in 1994, Sir Michael's report has had a profound impact on the industry. For example, it was pivotal to the changes in practice and legislation that made adjudication an accepted alternative to court battles. It also introduced the concept of collaborative working to an industry better known, historically, for conflict rather than co-operation.

"I'm often asked what I would change were I writing it now," he says, "but I'd change very little. I would put more emphasis on partnering, however. It actually features quite briefly in the report. But at the time, since no one had any experience at all of this way of working, the great thing was to get people to accept that there were other ways of doing things, so I concentrated on that. I would say much more now."

Although there are still many opportunities for the industry to improve, the report marks a significant change in what is perceived as possible. Sir Michael says he's often asked for his opinion on how things have gone.

"My answer is always the same," he says. "It's not gone as well as I hoped, but it's gone far better than I expected. It's a slow-moving and fragmented industry, but in my view there have been significant improvements in procurement approaches, particularly in the public sector, where I actually thought change very unlikely. Local authorities and health trusts are using best practice, framework contractors and, sometimes, have also insisted on a proper choice of subcontractors by main contractors."

This last point is very important. As Sir Michael says, best practice doesn't mean much if open book partnering contracts are underpinned by main contractors selecting subcontractors on lowest price.

"I continually say to clients, it's up to you to insist when choosing contractors that you get an assurance that they understand what partnering is about, that they're trained up for it and that they have chosen their own subcontractors on a partnering basis. The subcontractors have to be geared up as well. We've got to get the supply chain involved."

He's also aware of the particular problems involved in communicating the message to small and occasional clients.

"In numbers, although not in volume, they commission most," he says. "It's not easy to find a way of making them aware of the best way to go about that. We need to think about the places where they actually go, such as local chambers of commerce. When a small client needs, say, a new factory unit, they'll ring up an architect they know from the golf course. Now, the last thing that architect is going to tell them is that what they actually should do is approach a design build contractor because what they need is essentially a large shed. So getting the message over to these clients isn't easy at all, although bodies such as Constructing Excellence and the CIOB are making valuable efforts to do so."

Sir Michael is particularly gratified by the number of opportunities the report has generated for him to give seminars to clients, especially in local government, on the way to implement best practice approaches.

"There's interest now which just wasn't there 15 years ago," he says. "People used to go for lowest price and then wondered why there was a different result in the end."

In fact, Sir Michael remembers a key conversation with Sir John Bourn, auditor general of the National Audit Office, where he got the opportunity to put this idea across.

"I explained that there are two tenders," he says. "One's 'lowest price' and the other is the price with all the claims and variations added in. I remember Sir John making a note – "two tenders" – and I think that subsequent reports from the NAO show that the idea of best value was firmly adopted."

This increasing interest in best practice from the public sector has delighted Sir Michael. The figures from two NAO reports in 2001 and {2005} show that in the four years concerned, 700 million pounds of public money was saved, and the potential for two and half billion in savings would have been feasible had best practice been adopted across the board by Central Government Departments or Agencies.

"It goes to show what best practice can achieve," he says. "I'm concerned that the present economic climate will tempt some stakeholders to move back to the old days, but when times are hard, best practice is at its most crucial to successful business."

Sir Michael is well placed to comment – he has a macro view of the industry that stems not only from his years of experience, but also from a range of strategic appointments. He has many current roles, including: Chair of the CITB; Chair of Construction Skills; Chairman of the Joint Industry Board of the Electrical Contracting Industry; President of the Flat Roofing Alliance; Deputy Chair of Willmot Dixon and their housing company Inspace; and Honorary Fellowships at numerous institutions, including the CIOB and the Institution of Civil Engineers.

"I'm very glad to be able to have so many roles," he says. "It gives me a holistic view of what's going on in the industry. It keeps me out of silos, because I'm involved in so many organisations in so many different capacities."

{1960}

Euromast in Rotterdam, The Netherlands is completed.

{1963}

Kobe Port Tower in Japan is completed.

{1976}

CN Tower in Toronto, Ontario, Canada is completed.

{1976}

OCBC Centre in Singapore is completed.

Local school children | South Africa

Spencer Hodgson

{1943–   }

Fellow of the Chartered Institute of Building

While many of the leaders in this book have been responsible for landmark constructions, others are notable for their strategic leadership and influence on the industry.

However, Spencer Hodgson's career and life story is of particular note because it reflects the political development of a nation. Once a political exile, he went on to become an important figure in the reconstruction of his native South Africa.

Spencer was born into a family of white revolutionaries: Jack and Rica Hodgson were notable figures in the struggle for democracy in South Africa. His father, along with Nelson Mandela and dozens more, was arrested and charged in the 1956 Treason Trial. His mother was also detained in prison in Johannesburg during the {1960} state of emergency. Both parents, therefore, chose to undergo significant privations and dangers rather than accept the status quo.

"My wife says I was lucky to have been born into such a progressive family," says Spencer. "The chances of me supporting the system were always minimal. I didn't have to struggle to find my values, because I was brought up to believe apartheid was wrong."

Having thoroughly absorbed the ethos of his family, Spencer soon became interested in using his skills to create a fairer society. Keen on drawing and painting, Spencer was pleased to hear a career advisor's verdict that he didn't lack ability. However, recognising Spencer's inherent laziness, the same advisor suggested he'd better settle for architecture, rather than attempt "a really demanding course" at university.

Even so, any youthful indolence was soon erased by Spencer's growing desire to link his architectural skills to his beliefs.

"I wanted to understand what it would mean to create a just and equitable built environment," he says. "I still believe in working towards an industry that provides value to clients and society, but is accessible and creates more opportunities for employees as well."

In {1963}, Spencer's parents fled South Africa, making an illegal border crossing and travelling to England.

"I stayed on in South Africa until I'd matriculated and then went to join them," he says.

Denied a passport for refusing to join the army, he had to leave the country on an exit visa. He had no right of return, and would not see his homeland again for nearly three decades.

On completion of his studies, Spencer started working for a housing association in London.

"I got a lot of joy out of converting historic buildings in Camden, helping poorer people to get access to better housing," he says. "Right from my student days, I remember discussing how we were going to address housing challenges in South Africa. I don't think our understanding went much further than housing then, but we were already aware of the importance of the built environment to the country's future."

Much as Spencer enjoyed his work in London, he was not destined to remain there. After the {1976} Soweto uprising, the African National Congress (ANC) set up a school in Tanzania to receive pupils leaving South Africa. Spencer joined the team responsible for building the school, as well as a hospital, roads and bridges, factories, housing and farm buildings.

"I knew the project manager in Tanzania," he says. "Like me, he was a member of the ANC. He wrote to me and I decided to go. We had a workforce of about 400 people, but resource limitations. It was pretty labour intensive – for instance, we had to make all the roof trusses, doors, windows,

{1991}

One Canada Square in London becomes the tallest building in England.

{1991}

The new Stansted Airport terminal building in Essex, England is completed.

cupboards and furniture, so we established a carpentry workshop. We had a limited budget, so we had to attract funders. Ultimately, there was 100 million US dollars' worth of construction investment there. We had to make everything – only things like ironmongery were imported. We put in roads, sewerage and water. We built a dairy and a piggery. We were building a society, not just a school. It was an exciting time."

The Scandinavian countries had one of the most progressive aid programmes of the time, and were committed to anti-apartheid funding. The Norwegian First Secretary visited the project.

"He came in the early stages, and we showed him where we were going to build these things," says Spencer. "He came back a couple of years later and said: 'I've been involved in aid work for a while, and when you showed me your plans, I thought I'd heard it all before! But you have built it.' The thing is, once the momentum gets going and people believe in a vision, things happen. The project was really about providing an alternative, a model for what we'd like to see in the new South Africa."

The project was not without sacrifices for Spencer, however. He had malaria twice, and had to see his daughter suffer with it on several occasions. However, the goal – showing what might be achieved in a new South Africa – made it worthwhile for the family.

It also gave Spencer an opportunity to learn about collaborative construction.

"I picked up the ANC tradition of listening and taking decisions in a process that involved people – in this case, those involved in the construction and its users," he says. "You put up an argument and there's discussion and debate. You have to be persuasive if you want to win people over, and get the right decisions taken, without creating bad feeling. That was very helpful for me when I subsequently returned to South Africa."

His home-coming happened in {1991}, after an exile of 27 years.

"I didn't know the country any more," he says. "I had to relearn. I got a job with an organisation mobilising housing for the urban poor, in the townships around Durban. That was very rewarding, and in a way similar to my early days in community housing. It was a period of turmoil in South Africa, and there was a lot of violence. One colleague was killed, and we were once advised to leave a meeting by a different route to the one we'd planned, because people were waiting to hijack us. We were there to build houses or to create serviced sites for people to build themselves. Under the circumstances, managing all the dynamics was not easy, so I learned a lot there as well. Sometimes, we were helping people to get ownership of land they'd squatted for decades. They could then invest because they weren't going to get thrown off."

In 1996, under a new democratic government, he became a director in the Department of Public Works.

"I was involved in shaping the role of the construction industry in South Africa's development agenda," he says. "In 1996, when I landed in public works, many policies were still in the melting pot – we had to understand the issues at a macro level. In terms of the vision, our aims are fairly similar to those of other improvement agendas, such as those inspired by the Latham and Egan reports. We have to deliver value to clients and to society. But we're also dealing with additional social and economic issues that come out of our country's history – specifically, the empowerment of black people and black-owned enterprises."

In 1997, early in his career in public life, Spencer attended a one month intensive programme in Stellenbosch to help him gain a macro perspective. He has since sent a number of people, including the first women and first black woman to attend, so that they can have the same opportunities.

"I'm very proud to have played a role in the development of people like Bridgette Gasa, the first black woman president of CIOB Africa," he says.

Like many leaders, Spencer is acutely aware of the need to develop the leaders of the next generation, and has long been involved in helping young people to progress.

"When I was in Tanzania," he says "I also served on the ANC scholarship committee, sending students around the world. Some of these young people went away because they were interested in architecture and engineering. Some came back to take over work on the same project – they went on to build an ANC Development Centre near the Solomon Mahlangu Freedom College. Many of them have now gone on to become leaders in their own right in South Africa."

The inequalities under apartheid were so extensive that one black owner of a company reminded Spencer recently that he was arrested in the late 1980's for carrying a trowel in the street – black South Africans were not supposed to be skilled. In this context, both the enormity of the challenges facing the country, and the extent to which progress has already been made, becomes very apparent.

To overcome resistance from vested interests in the industry, Spencer worked with a ministerial task team (chaired by Murray and Roberts' CEO Brian Bruce) to draft important industry policy, Creating an Enabling Environment for Reconstruction, Growth and Development in the Construction Industry – and subsequent legislation. He then decided to apply for a job that would involve him in the implementation of the Act: CEO of the Construction Industry Development Board (CIDB) – a position he took up in 2001.

Having built the organisation from scratch, Spencer and his team were soon leading a complete reform of public sector procurement. To standardise the process and reduce confusion, only four contracts (FIDIC, NEC and two local contracts) were to be used. The CIDB also introduced a graded database of registered contractors, showing which firms could bid for which types of work and what size of contract they were equipped to deal with. Public projects were also registered, to ensure that contracts were not awarded by word of mouth to established companies. Affirmative procurement policies, promoting historically disadvantaged firms, formed part of the CIDB process.

"Leadership's a funny thing," says Spencer, "because there was a lot of resistance to this. The databases, which can now talk to each other, created an unprecedented level of governance and accountability. But in terms of achieving a more representative industry, particularly at the level of the smaller firms, we know it's working."

Currently, everyone is conscious that South Africa's construction industry has to deliver the infrastructure for the 2010 World Cup, but the far more fundamental challenge is to develop an infrastructure that meets everyone's basic needs.

"We're very aware that, as a developing country, our ability to grow the economy is closely linked to our ability to put the infrastructure in place. That's why construction is so crucial," Spencer says.

In 2007, he was invited to become a ministerial advisor. A year later Brian Bruce requested he join Murray and Roberts to support the company's contribution to skills development. The company has seconded him to the University of the Witwatersrand, where he is helping with campus redevelopment. The last building on campus was completed in the eighties, and about £100 million of construction is planned for the next few years.

"I'm very lucky to be working with a wonderful young engineer who heads up the group," says Spencer. "We're learning from each other. I just hope he learns as much from me as I'm learning from him! It takes me back to my days in Tanzania, when we had a vision about what kind of project to put together."

This time, Spencer's bold thinking is taking place right in the centre of Johannesburg. Naturally, there are many challenges to overcome, including funding constraints and the need to integrate a somewhat fragmented campus.

"Part of mobilising this process is ensuring that we adopt the best practices that I was part of developing in my previous job," says Spencer. We're introducing NEC – with great pain, I must say, as it's a complete culture shift. So we are on a learning curve together with our consultants and contractors. In the long run, it will pay off."

Despite the challenges, he's thoroughly enjoying his work. After a lifetime devoted to improving the built environment as a means of improving society, he is delighted to see how far things have progressed.

"This is a wonderful university with proud moments of opposition to apartheid," he says. "I walk around and think about how far our country's come. It used to be exclusively for whites, but now we have 70% black students and 50% women. I'm working for education, just like I was thirty years ago, so in a sense I feel like I've come full circle in my career. At the same time, so much has changed for the better."

Sir Joseph Dwyer

{1939–   }

President (1998–99) of the Chartered Institute of Building

Sir Joseph Dwyer, the man best remembered for the Channel Tunnel and his canny asset-swap with Tarmac, started his career in construction at 16, when he became a chain boy (i.e. surveying assistant) for Wimpey.

The future chairman of that company could very easily have been lost to the construction industry, since he was also a talented young sportsman. As well as playing county-standard table tennis, he was a high-level amateur footballer too. He remembers playing matches at Liverpool's Anfield stadium and Everton's Goodison Park, and he also had a trial for Crewe Alexandria in the mid-fifties.

A dual career seemed fleetingly possible to the young Joe Dwyer, but the construction industry worked a six day week back then. Getting to Crewe for the football on a Saturday afternoon was a complicated business, and inevitably, his absences from work were discovered. His manager told him to choose between studying for a career in civil engineering and playing football.

> ### "At that time, £20 week was the maximum wage for footballers," he says, "so I decided to stay with the day job. These days, things might have been different!"

He progressed quickly once fully committed to a career in construction, becoming a site manager

by the age of 24. He put in many years at night school, and his continuing hard work allowed him to qualify as an engineer at what is now Liverpool John Moores University.

By his early thirties, Sir Joe was Regional Director of Wimpey in the north west of England. He is keen to point out that this early start in the world of work, which is not generally available to talented young people today, enabled him to amass experience and to progress through the ranks at a reasonably early age.

It wasn't easy, though – the early eighties brought recession, and he had to find work for almost 100 permanent staff in a tough economic climate. The housing market was tightening and the pressure was on. Long before the famous asset-swap with Tarmac, Sir Joe was exercising his leadership skills through testing times.

One major contract, to rebuild Merseyside's Cammell Laird shipyard, fell to Wimpey after the original contractor pulled out due to problems with the unions. A group of workers locked themselves into the yard as part of an industrial dispute, and negotiation proved fruitless. A difficult stand-off followed, which was eventually resolved by driving Wimpey's construction plant through the locked gates. However, this firm action was ameliorated by Sir Joe's desire to give work to as many of the strikers as possible once the period of militancy was over.

The incident confirmed Sir Joe's growing reputation, and he was earmarked for promotion.

"We used to have regional organisations dealing with every type of business, from housebuilding to quarrying," says Sir Joe. "When the structure of the business changed, I was asked to form an office in Manchester for housing and construction."

By the mid-eighties, Sir Joe had moved up again, becoming Wimpey's MD for Construction across the UK. He spent a year living out of London's Pall Mall club before his family were able to move south. He also spent some time as director of Wimpey Homes and Chairman of the Minerals Division – his involvement at a senior level in so many aspects of the company making for a busy life.

{1995}

San Francisco Museum of
Modern Art, designed by
Mario Botta opens to the
public.

"We built quarries in Bohemia
and many airports, including in
Slovakia," he remembers. "We also
built an airport in Tibet, bringing
in equipment by rail. There were
Russians with Kalashnikovs posted
to stop pirates."

Although Sir Joe did not realise it at the time, he was being
groomed to take over the entire business. He was given experience
of projects from Canada to Dubai, and became conversant in all
aspects of the business.

So, what does Sir Joe think marked him out to lead the company? It's
not an easy question, but he does remember one insight he gained.

"We were given psychometric tests on a couple of occasions," he
recalls, "and I was told that most people are either good at strategy
or good at detail. Apparently, I'm unusual because I'm good at
both. Also, I'm generally a very placid person. I don't get emotive.
That helps. I'm very pragmatic, although that means I can't do
cryptic crosswords!"

Sir Joe's next strategic appointment was to the supervisory board
of the Channel Tunnel. As the then chairman of the contractors'
group, his robust negotiations with the hard-driving client
representative, Sir Alistair Morton, became the stuff of legend.
Eventually, the monumental time required for this role forced him
to move back to his day job at Wimpey, and he was succeeded by
Neville Simm of Tarmac.

Perhaps Sir Joe's greatest project, the Channel Tunnel was widely
discussed in the media, but he finds it sad that the journalists were
more interested in reporting the commercial arguments than the
project itself. The challenges of his relationship with Eurotunnel's
Sir Alistair Morton were comprehensively documented, but the
wider public heard little of the technical achievements.

Of more interest to Sir Joseph were the learning opportunities
afforded by this international construction project, particularly the
differences between the French and British in terms of method.

"For instance," he explains, "when we extracted the chalk, on
conveyor belts a quarter of a mile long, we tipped the spoil into the
Channel. Using the same machines, the French tipped the spoil
into a mixing chamber, suspended it in water and pumped it

overland to an inland lagoon. They then re-circulated the water. It
was a much more technical solution, but it cost no more to do."

The following year would see the construction of another of
Sir Joe's greatest projects, the concrete oil drilling platform of
Hibernia Field, Canada. At a million tonnes deadweight, it was the
largest in the world at the time.

"It was built to withstand passing icebergs," he says. "As well as the
technical challenges, the Canadian government required that we
employ the local fishermen on the project. There were moments of
great difficulty – but it worked out okay in the end."

The phase that Sir Joe remembers best was the completion of the
huge concrete structure, which had been built in dry dock with an
earth wall holding back the sea.

"It was moved with the aid of buoyancy jackets," he recalls. "The
earth wall was broken and the Atlantic rushed in over 176
thousand tons of concrete and steel, which slowly rose to the top.
The superstructure was built while it was in the water. Then it
was towed out to its final resting place and filled with ballast. The
process was awesome to watch."

As the tunnel and oil drilling platform in Canada was completed
in the early nineties, recession was once more biting the economy.

"In recession," says Sir Joe, "investment stops and you try to reduce
borrowings. The only way to do it is either by raising share capital,
which is hard in a recession, or by selling assets."

As Chief Executive of Wimpey, Sir Joe sold off development
property and eliminated borrowings. As the recession eased in
{1995}, his thoughts turned once more to investments.

"I'd been Chief Executive for three or four years," says Sir Joe,
"and I was committed to safeguarding the balance sheets of the
company. The job really brought home to me an understanding
of conglomerate businesses. The received wisdom is that they are
synergistic, since different parts of the business do well at different
times. I'd said it before, as did my predecessor, but I was starting
to doubt it."

The contracting division was not making any money. In a
competitive market, Sir Joe realised that people were bidding for
work purely for turnover. Profits were slim to non-existent. It was
time for a rethink – on several levels.

Channel Tunnel | United Kingdom

{1999}

Burj al Arab in Dubai,
United Arab Emirates is
completed and opened.

{1999}

The London Eye opens to
the public .

{1999}

Millennium Dome
in London, designed
by Richard Rogers, is
completed.

{1999}

Jubilee Line Extension of the
London Underground Jubilee
line is completed, featuring
innovative and widely
acclaimed station design.

"I decided that if we were getting out of contracting, we needed to raise share capital to move forward," says Sir Joe.

This led to one of the many delicate business negotiations of Sir Joe's career. At the time, a family trust was the majority shareholder. There was a concern that if the trust failed to take up its shares, which it was reluctant to do because of an unwillingness to dilute the existing holding, then the share issue would fail. Luckily, after some delays, the trust decided to sell all of its interest in the business, and Wimpey became a full Public Limited Company.

By now, Sir Joe's rationalisation of the business had left it with housing, contracting and minerals divisions. The mineral assets, including several limestone quarries, were valuable because there was little chance of further quarrying activity being given planning permission in the UK. However, they were not making a profit; 80% of the firm's profits, he realised, were coming from the housing division.

Just as Sir Joe was reviewing the business, Neville Simms put Tarmac's housing division (McLean Homes) on the market. Conversations with Simms revealed that he had a strong interest in the mining sector and Sir Joe saw an opportunity. He suggested they do a swap – McLean Homes for the Wimpey quarrying division. This led to weeks of talks, and several times it seemed that the deal would not go ahead – largely because McLean Homes showed more profit than the Wimpey quarrying business. Eventually, Sir Joe offered Tarmac the Wimpey contracting division as well, and undertook to retain all risks associated with the Channel Tunnel concession. Finally, the deal was completed in March 1996.

As the nineties drew to a close, Sir Joe's thoughts turned to retirement. After the asset swap, he'd always reckoned on spending three years getting the new structure into shape and making sure everything was properly integrated. Then, he felt it was time to go.

"They asked me to stay as a non-executive chairman," he says, "but I refused because I felt everyone would have continued to defer to me."

He didn't see this as fair to his successor, so he started making arrangements to step down.

During his final year at Wimpey, he became President of the CIOB. Uniquely, he has been president of the CIOB and of the Institution of Civil Engineers (in 2001 – when he also received his knighthood).

"I'm proud to be called both a builder and a civil engineer," he says.

On June 30th {1999}, at the end of his Presidential term at the CIOB, he had a memorable day.

"I retired from George Wimpey after 44 years," he says. "It was my sixtieth birthday. The same day, I attended the CIOB's AGM to complete my term of office and hand over to Paul Shepherd. Then I flew to Manchester, got in the car and returned to Liverpool to launch the first urban regeneration company!"

Urban regeneration schemes, he remembers, were created in response to a report by Richard Rogers on regeneration, and several urban areas were selected to pilot the idea.

"I wasn't sure when I was approached," he says. "It was a period of change in local government, because the regional development agencies had just been created. But I got the impression that there was a real opportunity to achieve something, so I said yes. From then on, I was chairman of Liverpool Vision, working with public sector partnerships."

However, when the organisational structure of the regeneration effort altered in {2008}, Sir Joe stepped down from his role to become a non-executive director at Crossrail. Sir Joe – who is also a governor of Liverpool John Moores University – remains pleased by the changes made to the city centre and waterfront during his tenure of Liverpool Vision.

"We had nine years of reasonable success attracting £3 billion worth of investment, and Liverpool has now become an international tourist destination," he concludes.

Half a century after joining Wimpey in the North West, it seems that Sir Joe is still focused on the region's buildings – and on building the region.

He has been honoured with Honorary Degrees from the University of Liverpool, Liverpool John Moores University and Nottingham University. He has also been commissioned as a Deputy Lord Lieutenant of the County of Merseyside, and finally the accolade of "Citizen of Honour" from his home city of Liverpool.

His wife, Lady Stella continues to ask him when is he going to 'really' retire – alas to no avail.

Channel Tunnel and Eurostar | United Kingdom

Olympic Stadium (Montreal) | Quebec, Canada

Professor Roger Flanagan

{1944–   }

President (2006–2007) of the Chartered Institute of Building

Roger Flanagan's father, in common with many parents of the time, had hopes that his son would be able to take advantage of the opportunities offered by the post-war world to build a better life.

"He wanted me to join the professions, no matter what," says Roger. "He never had a lot of money, and he always said he didn't want me to have to struggle like he'd had to."

When he talked about the professions, however, it was more or less taken for granted that he would join a construction-related profession – the only question was which one. Roger's father was a bricklayer and small contractor who quickly communicated his love of the industry to his son.

"I always thought it was a fabulous idea as a young boy," he says. "I'd say to any young person – I've never regretted joining the industry. I got the bug from my Dad, and it's brought me travel, excitement and work at the boundaries of technology – it's a passport to anything."

Mindful of his father's advice about professionalism, he went to study at a forerunner to Surrey University. He remembers his younger self as being a "hungry young Turk", who took on a variety of jobs after graduation. For a while he was involved in precast concrete housing, but one of his most memorable jobs was working on the Montreal Olympic stadium.

The politically sensitive project had gone hugely over-budget, so Roger was recruited to help gain control of the costs, commuting to Canada every month.

"The Canadian prime minister had to sanction expenditure on the job at one stage," he remembers. "I also worked on logistics for a hospital project at Pamplona in Spain."

International working, so much a feature of his later career, was already becoming a habit. When he visited the United States in the 1970s, Roger was strongly influenced by the speed of construction, use of offsite construction and standardisation of services.

"At the time, they were building faster than anyone else in the world," he says.

"I'd already become totally hooked on understanding better ways of doing things. I'm someone who very much believes in moving on and moving up, always."

In a way, for Roger, this was a reaction to the face of construction he'd experienced as a boy.

"I watched my father," he says, "and he had a tough life. He was in pain from a slipped disc, and was eventually bankrupted by a single client who walked away without paying his bills. It's totally brutal – he was an honest bloke trying to do an honest job, but the client didn't pay. It wasn't about the quality of the work, it's just how they behaved. There wasn't a lot a small guy could do. He never really recovered from that, and died when he was only in his early fifties."

An innately optimistic person, Roger turned this sad episode into a positive influence, convinced that finding different ways of thinking and doing things were the only way to improve the industry. He took a Masters Degree at Aston University.

"Aston was a great place to go," he says. "It had a great reputation and a good track record in engineering. It was an absolute revelation. We were at the leading edge of a fast-moving industry. Afterwards, I knew that I wanted to focus my career on the international market. I did a number of assignments in various places, but also went into lecturing. I wanted to span industry and academia."

Roger is at pains to stress the importance of industry links if academic courses are to achieve their potential for the construction managers of tomorrow.

"Having been involved in projects all over the world, I can talk to students about the real issues," he says. "For example, how do you keep a workforce of 4500 managed, motivated, fed and

{1983}

Burrell Collection
Building in Glasgow,
Scotland is completed.

{1983}

Trump Tower in New York,
United States is completed.

{1983}

Wells Fargo Bank Plaza in
Houston, Texas, United
States is completed.

{1985}

The first tower in the
World Financial Center
in Manhattan, New York,
United States is completed.

up to speed? This isn't about contracts and claims – it's about decent working conditions and decent food. I remember a project where the workers threatened to strike because the food wasn't hot enough – this is the real world our students will be working in. I passionately believe that if you're going to teach the next generation, you have to have had some involvement in the industry."

The nature of Roger's working life allows him to gather perspectives on different segments of the industry on every continent.

"I work on projects," he explains, "but they don't consume years of my life. A lot of people hire me to create a snapshot on progress, at a particular moment in time."

The snapshot might be of a project, but it might equally be of a national industry. Organisations such as development banks will hire Roger to explore what's shaping a nation's progress in construction and what the industry might look like in the future.

Major studies of the Japanese and Chinese industries were memorable commissions from the 1980s.

"I first went to China in {1985}," says Roger, "and I've watched the country develop ever since. Then, it was a centrally planned economy with the door ajar – now it's wide open. I've witnessed a fantastic transition. Similarly, I first went to South Africa in {1983}, and I've watched it grow and become a really exciting force for change within Africa."

Roger was also involved for 17 years in the Davos-based World Economic Forum, as one of the governors for the now-disbanded Engineering and Construction section.

"Being involved with that group for so long had a huge influence on my thinking," he says. "One could only go by invitation, and it gave me access to the top 50 engineering and construction firms in the world. They would come to Switzerland every year to shares their views and visions, which gave me a huge insight into how trends were developing."

As well as accruing several visiting professorships over the course of his career, Roger has served on the board of companies as far apart as Switzerland, the USA, Hong Kong, Sweden and the UK.

Notably, he served on the board of Skanska as a non-executive director for 10 years (their maximum term), and was the first person from outside Sweden to do so. He gained valuable insights watching an expansion that spanned Eastern Europe, while at the same time the firm acquired the construction division of Kvaerner.

Through all this, the one constant has been his work at Reading University, where he is Professor of Construction Management.

"It's the most fabulous place to be," he says. "Whilst I travel a lot, Reading is my bedrock. I'm 100% engaged in teaching students. They're the reason for being there – they are tomorrow. I'm not being altruistic – I recognise we need to produce the best and most forward thinking people to tackle the challenges of the future. A lot of people don't understand why teaching's so exciting, but there's nothing I would rather have done than this. As you build a relationship with your students, you become part of their career path. It's not the dull job people imagine. When I go and watch my students graduate, knowing that they'll be successful, it's such a buzz. For me, it's like driving a Ferrari."

Many of Roger's former pupils are now in top positions around the world, including one student who is Vietnam's national deputy for construction. Roger can pick up the phone and talk to senior industry figures in dozens of countries, because they all remember their time at Reading with affection. Notably, he stays in regular contact with Li Shirong, the first woman to become president of the CIOB.

Roger was himself President in {2006}–07, which he describes as a fantastic experience.

"I was thrilled to be considered – it was a privilege," he says. "I met people from all levels of the industry. There's a perception that the bottom end of the industry is weak, but I can honestly say that all I experienced at every level was professionalism. Our SMEs often feel that they're at the bottom end of the food chain, getting all the bureaucracy piled onto them, but they come up trumps time and again through their sheer professionalism."

The CIOB's insistence on standards has a particular resonance for Roger, when he remembers the difficulties his father faced.

"The CIOB gives out the crucial message that you can't step over the line because people will hold you accountable," he says. "Non-payment can cause problems that nobody should have to go through, so I will always be committed to the CIOB's insistence on professional standards."

{1985}

DLI 63 Building in Seoul,
South Korea is completed.

{1985}

HSBC Headquarters
Building in Hong Kong is
completed.

{2006}

Construction begins on
the Freedom Tower, a
replacement for the World
Trade Center.

Setting his own professional standards, Roger has always been active as a researcher and author, as well as reporting for major institutions such as the World Bank and teaching tomorrow's leaders.

"My early books were all about comparing national industries," he says. "I was trying to understand the underlying drivers that shape each country's industry to perform the way it does. The industry is very different around the world, and construction is a generic term spanning a lot of things. I'm very intrigued by Scandinavia. They've got 7.9 million people but some of the world's best companies. It's a cold and hostile environment with few natural resources, but they're damn good at building."

He's keen to stress how we need to take a step back to understand the dynamics that shape each national industry. For instance, he argues, the UK's reliance on brick can be understood in terms of the plentiful supply of clay, which can be transported in a way that would be almost impossible in, say, Australia.

To date, he's authored or co-authored 14 books, encompassing topics as diverse as risk management and whole life costing.

"People talk about managing risk and have all the right bits of paper," he says, "but in reality risk tends to be monitored rather than managed. That's where it goes wrong. I feel strongly that this is a most important area. When you work on a 2–3% margin, that in itself creates major risk. You only need a dollar fluctuation to wipe out your profit. The problem isn't the margin itself, it's the risk. In our everyday lives we have illusions of certainty. Just travelling to a meeting, a hundred things could go wrong, but we still arrange our lives as if we could guarantee where we'll be and when. But it's a delusion!"

His work is thought provoking and covers many diverse fields, but all of it is informed by a wider desire to improve the status of construction.

"We need to develop an image that puts us on a par with the major professions, from architecture to dentistry," he says.

"We had to fight to get construction into the universities – there was always a striving to be accepted. The industry improved, but we lacked a strong intellectual base and respectability. I think we've got that now. We've reached a point now where we've matured, come of age. Now academia needs to go back and engage with the industry again."

Roger has watched the construction industries of the world mature and develop, but he's still a big believer in the strengths of his native UK.

"I honestly believe we've got some fascinating assets, but we're so good at talking ourselves down," he says. "I think it's cultural. But if you say you're lousy for long enough, people believe you. That's what frustrates me about the whole UK industry. We never bang our own drum, but there isn't any country that can do better. There's nothing technically that would phase a UK firm – what we lack is belief and confidence."

As evidenced by his top-flight pupils now working across the globe, he believes that graduation from a UK university still opens doors around the rest of the world. He's equally keen to share a positive message with young construction professionals in the UK.

"The future is not bleak, it's just different," he says.

"We need to motivate and encourage young people, to show them the window of what's possible. Throughout my career, I've been trying to communicate a vision."

The day we speak, Roger has just taught a class of 36 students, representing 17 different nationalities.

"I can't tell you what a privilege it is," he says. "They're so keen to learn, and I'm teaching the people who will shape all our futures. It's a fantastic experience."

Yi Jun

{1960– }

Fellow of the Chartered Institute of Building

Yi Jun, president both of China's foremost international contractor (CSCEC) and the CIOB in China, had a patriotic reason for applying to study civil engineering as a teenager.

"I saw it as something I could do for my country," he says. He studied for four years and found the university experience immensely rewarding. In common with many Chinese students, Jun developed a great appreciation for the way he'd been helped by those who mentored and taught him during those years.

"I respected my teachers very much," he says. "Without exception, they were all very kind and helpful to their students. They taught me so much."

Graduating with a Bachelor of Science in Civil Engineering from Chongqing Construction University, he was completely committed to creating a successful career in his chosen field.

"When I joined CSCEC after my graduation in 1982, the work experience only intensified my love for this career," he says.

Jun's first role was as a technician on the Lidu Hotel Project, but it wasn't long before he had been moved to the Business Operation Division of the company. Within two years, he was working as Section Chief on the Shenzhen Lianjian Tower Project. Shenzhen is one of the great new cities of China, built very quickly over the past few decades. The acceleration of Jun's career was to be equally rapid.

He was returned to CSCEC's training centre to learn English and to take a course in bid management, and continued to be enthusiastic for opportunities to learn.

"I studied construction project management at Cardiff University in the UK between {1986} and {1987}, and I achieved a Masters degree in engineering from Chongqing University in 1999," he says.

While in Cardiff, Jun trained for 18 months at Taylor-Woodrow, working in many different departments of the company. Returning to China, he was made project manager on the Beijing Lufthansa Centre.

"This was a much bigger challenge," he says. "It was a massive international joint venture, with German partners, so there were lots of complex issues to deal with. To begin with, there were basic issues of understanding which had to be overcome. That's not just because it was, of course, a multi-lingual project. There were different methods of costing, for instance."

The challenge was indeed immense. The owners state that the completed building holds more than two dozen meeting and conference rooms, 120 offices, 160 modern apartments and more than 520 hotel rooms. There's also an exclusive shopping centre, bank branches, air ticket sales outlets, doctor's practices, sports studios, beauty salons and Montessori kindergartens.[1]

"It was also a multi-faceted project," says Jun. "Everything from a shopping centre to a hotel had to be constructed on a very tight schedule. That meant that there were time constraints, but I couldn't let that compromise on the quality. For the first time, I was project manager on a job which would be constructed to international standards and contracts. I found the experience inspirational. There was a bigger gap in my knowledge than I had expected, but that just inspired me to learn and change."

It wasn't just Jun's personal development that he felt he needed to examine. Based on his experiences from that project, he established a new enterprise: the Construction Contracting Company, where Jun was president from 1993 to 2001.

1 http://china.lufthansa.com/en/html/destinationen/vorort/index.php Accessed 11th February 2009

{1986}

Temasek Tower in
Singapore is completed.

{1986}

Robot Building in Bangkok,
Thailand is completed.

{1987}

One Atlantic Center in
Atlanta, Georgia, United
States is completed.

{2003}

Walt Disney Concert Hall,
designed by Frank Gehry,
opens in Los Angeles,
California.

"It took a whole year to convince my superiors to establish the new company," he says. "I had to sell my ideas on how important it was to the group and government. But I knew it was the right time. People eventually came to see how important construction management and contract management would be. To begin with, however, these new ideas about contract management were very difficult to accept. In terms of familiarity with the concept, I was starting from zero, so it was hard to convince the decision-makers. I had to make very careful, reasoned arguments to support it. But we did it, and my whole team was extremely proud to get the final agreement."

Already influential at a relatively young age, this was only the beginning for Jun. In his early thirties, as president of this offshoot company, he was jetting around the world setting up its major strategic deals. In the first seven years, 30 projects were won in countries including Japan, Finland and France.

At home, there was work on the German Embassy and the Sunflower Tower in Beijing.

"We got so many contracts because we had the linguistic ability, we had the core competencies, we had the investors," he says. "International contracting can be tough, but you find a way. It's always about solving problems."

Alongside the normal challenges of construction, Chinese businesses have had to adapt to new ways of thinking and working in the wake of reforms and the 'open door' policy.

"The major issues now for construction in China are the challenges deriving from pre-existing operational modes and capabilities," explains Jun. "There are also sometimes sub-optimal organisational structures, and a relatively slow rate of adoption of new thinking about management. And thirdly, changing markets mean varied demands from diverse investors, which really challenge traditional operational concepts and service methods."

Starting new companies, therefore, is only one of the ways CSCEC has needed to adapt in response to China's changing business environment.

"We have been actively carrying out reforms to respond to the challenges and needs of the changing markets which we've seen over the past decade," he says. "It's taken persistent efforts, but it has to be done to safeguard our core competitiveness. It's also about development strategies for the international markets. CSCEC has to optimise its structure and expand its business chain. We seek a breakthrough in huge, high-end, compound projects – put simply, we want 'big markets, big clients and big projects'. Our goal is to be the most competitive Chinese conglomerate construction and real estate development enterprise on the international market, to enter the top 300 of the Fortune Global 500 and to be among the top three global construction and real estate development conglomerates in 2010."

Developing the presence of CSCEC (USA) Ltd is also a big priority. Foreign contractors can find it as hard as foreign musicians to 'break America'. But the CSCEC subsidiary, started in 1987, had won more than 70 contracts by the end of 2006. The big breakthrough in the American market came in {2003}, when CSCEC (USA) Ltd was awarded the contract for a US$22 million middle school in South Carolina. The US$130 million contract to build the New York Marriott hotel followed at the end of that year.

Jun's impact ensured that by 2001 he was Vice President of China State Construction Engineering Corp., and Chairman and President of China State Construction International Co.

"I focused on restructuring, and ensuring that our practice and communications capabilities were truly international," he says. "We became able to take on major overseas projects. For instance, we built an international airport in Algeria. That was a big project!"

Big projects are typical for this evolving Chinese company. It has been working on what's planned to be the highest building in Europe – the Federation Tower. Designed by Norman Foster and Partners, this building will become the tallest in the world to use a natural ventilation system.

For Jun, however, there are some drawbacks to working on these mega-projects. "I'm always looking at the overview of a project," he says. "I don't have time to get involved with individuals or with the details on a project. In that sense, I miss being on site. I'm always involved with multiple projects, and constantly travelling. My work now is about establishing relationships at governmental level. I have to think about what's important for the whole company."

Jun is now president of CSCEC Ltd and chairman of the board of China State Construction International Co. Somehow, however, he also finds the time to be President of CIOB China and he's keen to stress the potential of the Institute's role in the global construction market.

"It's important historically, but it can also do much for the future," he says. "It plays a crucial role in the training of its members, no doubt its fundamental responsibility. This has to continue. At this time of global challenges, the CIOB needs to equip its members to lead the industry towards a better future.

We could also do more in the promotion of scientific building concepts and modes. Knowledge sharing across a global platform has to be the CIOB's most exciting and relevant offer for the future. The lure of sharing knowledge and information on market promotion will bring more construction elites to join. That's what will make us a truly global force."

Yi Jun

China Central Television (CCTV) Tower | Beijing, China

Jun also focuses his vision at a global level for his own corporation. Currently, CSCEC concentrates on the special administration areas of Hong Kong and Macao, as well as Singapore, Vietnam, Algeria, Equatorial Guinea and Botswana, the USA and Barbados, and the Middle East, particularly the Gulf. Jun has detailed plans.

"The competition in the international market remains very tough," he says. "We have to achieve on many levels – technical, cultural, managerial, contractual. It's not easy to become a significant global player. We pursue this goal by making sure we have the right staff, and incrementally developing our position in the market. We still have gaps in our skills, but we're closing them all the time. I'm currently looking to increase the ratio of infrastructure to real estate projects. A good example is the consultancy service contract we won for the new administrative city in the Republic of Equatorial Guinea, with a value of 621 million euros, which is the largest contract of its kind won by a Chinese enterprise to date. We also actively promote localisation in management and the effective use of local resources. Having said that, our business strength lies in our growing world-wide reputation for the highly qualified, professional teams deployed on our international operations."

One job that created a step-change in the company's growing profile was the Beijing Olympic Games. CSCEC was involved in the construction of no less than 18 venues, including the iconic Beijing Aquatics Centre (commonly known as the Water Cube). Interestingly, this building is not in fact a cube, but a rectangular box-shape. Even so, the seemingly-random pattern of bubbles that make the building so iconic has won it numerous awards.

"I was very proud of our involvement in the Games," says Jun.

"It was a huge technical challenge, having to produce world-class facilities in a very tight time-frame. The deadline was immoveable – we knew that the whole world was coming, and that we had to finish!"

The Olympics gave Chinese constructors a chance to display their capabilities to a global audience, and their achievements can only enhance their standing in the international market. And such a change seems likely to continue to typify the progress seen by Jun in his own working life.

"I have spent my whole career with CSCEC," he says. "That's 26 years so far. Becoming CEO was a surprise. I never expected it. I appreciate the support of the whole group as I'm doing my job. I work very hard, but I always appreciate and remember the contribution of the team. Becoming CEO, after all, is just chance, in a sense. Many would like the job, but there can only be one at a time! Why do I think I was chosen? I think because I give my total concentration to whatever I set out to achieve. A leader must have energy and passion. You have to try to inspire your colleagues. It's always about the whole team. If you can remember that, you can achieve most things."

Beijing National Aquatics Center 'Water Cube' | Beijing, China

Howard Shiplee

{1946–   }

Fellow of the Chartered Institute of Building

Like many young people of his generation, Howard Shiplee found that decisions about his future path came at an early stage.

The multi-tier educational system of the time streamed children from the age of 14, so this is when Howard found himself enrolling in a building faculty at a technical school.

In those days, there was more emphasis on streaming and less on individual career fulfilment, but Howard found that failure to reach grammar school was in fact the beginning of a lifetime of opportunity.

Now, as Construction Director at the Olympic Development Authority (ODA), he still remembers that pivotal moment and strives to give today's young people the same kind of chances to succeed.

"Going to a technical school was a turning point," he says. "Ever since, I've been in one of the best industries one could hope to work in. It's been good to me. I've had a wonderful career, working with great people from all over the world. I see the Olympics, and the scope it's given to develop training and best practice in the industry, as an opportunity to give back to society."

This is clearly very satisfying to Howard. There's a sense in which his career has come full circle, as the erstwhile apprentice takes a leading role in the ODA, a key driver in creating positive changes for the apprentices of the future. However, he is keen to stress that this is not about the execution of some carefully conceived master plan.

"People talk about career planning," he says, "but I've always found it very difficult to plan. I've moved around an awful lot – for me, it's all about finding opportunities."

As a young man, he had a number of jobs, studying hard to gain a degree at the same time. He has fond memories of working for Sir Alfred McAlpine's during his early career, but he nevertheless kept moving, anxious to get experience as fast as possible.

By the mid-seventies, when the industry was going through challenging times, it was not immediately clear what the next opportunity would be. Out of the blue, he got the offer of a job in the Middle East. This proved to be an experience that really opened his eyes to how the industry could work at its best.

"We were working with oil and gas companies that were very enlightened about developing relationships with their contractors," he explains. "They weren't interested in contractual wrangling – their prime imperative was always to get oil out for sale to their customers, and that had a huge influence on how they worked with their partners. It influenced me greatly – they were collaborative, but always robust and effective in their dealings."

Having been more used to the culture of claims and counter-claims, this was an exciting insight into another way of working.

"The claims culture at that time was bad, but not entirely industry's fault," he points out. "Clients are not always fair and equitable. But the more enlightened industries were showing how things could be done. From that time in the seventies, what changed is that clients finally woke up to the fact that when there are contractual

{1999}

Jewish Museum Berlin, designed by Daniel Libeskind is completed.

{1999}

British Museum is redesigned by Norman Foster.

{1999}

Reichstag building in Berlin is reopened.

{1999}

Melbourne Museum by architects Denton Corker Marshall is completed (the largest museum in the Southern Hemisphere).

arguments, they're paying the bill whatever happens. They realised that there was a better way to go."

Howard received more exposure to enlightened ways of working at Broadgate, where an American-style model was used to create a more collaborative working arrangement.

"This was the birth of the specialist contractor," says Howard. "As the main contractor became more involved in the management of specialists, the dynamic changed. It was possible to create a new environment that was very collaborative and very effective."

Another pivotal point in Howard's career came when he was approached by Amec and given the opportunity to go up the ladder into a more senior operational role. For six years, he was northern operations director for Amec projects, which gave him a significant stepping stone into larger developments. During the late eighties, most major companies developed a requirement for computer facilities as ways of working changed. Howard oversaw the development of major facilities for clients such as ICI and Barclays, always remaining focused on service provision through non-adversarial contracting.

Then opportunity came Howard's way once more, as he was invited by Amec to become Project Director on Hong Kong airport's new international terminal.

"I remember getting the call," he says. "The contract had just been awarded, and I was invited to go out, have a look and talk it over. So that was another step change in my career."

This was a major international joint venture and there was a big spotlight on the project. Sovereignty of Hong Kong changed halfway through the build and Howard describes the contractual arrangements as "very robust". Nevertheless, he was able to build a good relationship with the client-side representative for the Airport Authority – Douglas Oakervee, who went on to become president of the Institution of Civil Engineers (ICE).

Despite the challenges of the contract, Howard did not lose sight of his core values.

"Historically, the Hong Kong construction industry's health and safety culture hadn't always been good, so we worked really hard on safety," he says. "We and the client wanted to make a difference. And in 34 million man hours, we had no fatalities. At the time, that was unheard of there – and we reduced the incidence of accidents to one third of the norm."

Howard is pleased by this, not just because people were safe, but because it reflects on the efficiency, planning and commitment of the contractors.

Problems on the project were solved collaboratively, ensuring a satisfied client and a reasonable return for the project partners.

"We had real issues," says Howard. "It was a high profile environment with a complex set of relationships and yet we had to get the job done in record time. I think it's probably one of the fastest-ever constructions of a major international terminal."

The team were faced with major technical issues – it was a challenging structure – so they set aside all contractual matters, put the best people together from all sides, and instructed them to find the most economic technical solutions.

"I wanted people to contribute in a way that was unfettered by the contractual position, so they could concentrate on obtaining the right technical answer," says Howard. "Only after that would we have a discussion about roles and, ultimately, where the bill might fall. But as soon as we'd got a technical solution, I'd want the same level of rigorous attention to the contract because we knew we might have to spend a lot on implementation."

Achieving such an efficient and collaborative approach sounds almost utopian, but according to Howard the answer is simple.

"There is a level of trust in industry," he says. "There always has been. It's a very personal industry. There are a few key players on both sides who make the running. You can get everyone who needs to be involved together in a big room. That's what builds the trust, and we absolutely have to have it in construction, because everything we do is a prototype. Everyone just has to do their best and trust the clients to pay. The clients have to trust us to deliver their capital facilities through a process which is very unfamiliar to them. I'm not advocating naivety – there are many things which it is appropriate to get in writing – but it causes the industry huge problems if we can't build reciprocal trust."

At the end of the airport project in {1999}, Howard returned to the UK and worked for Capital Land, which manages investments for Singapore's sovereign fund. Howard joined the board to advise and control delivery of London-based projects, getting involved with the Canary Wharf development, which he describes as another big operation from the enlightened end of the market.

{2004}

Scottish Parliament
Building by Enric Miralles
is opened.

{2004}

30 St Mary Axe, London
(the Swiss Re building), is
completed. It symbolised the
new high-rise construction
boom in London.

"Projects were managed in a way that got them to the marketplace quickly and efficiently with joint contracting partners," he says, "and it was interesting to observe."

It wasn't long, however, before opportunity knocked once more. Howard was invited to become director of Carillion Building, with particular responsibility for PFI delivery. He spent two years there, leading the development of the Government Communications Head Quarters at Cheltenham. He says it was a large and complex project, delivered for a very sophisticated client, but beyond that he's not at liberty to comment.

Late in {2004}, he joined Railtrack's Thameslink 2000 programme to upgrade the north / south route of London's rail network. He expected to commence work when he joined, but it very quickly became apparent that it was not going to go ahead as expected.

"This was a disappointment," says Howard, "because I'd been looking forward to learning within a new technical environment, but the time provided me with an opportunity to gain a deeper understanding of railways. Rather than getting straight out to build bridges and overpasses, I had time to work on refinements to things like the signalling system for busy intersections. I could also look at the totality of the operational parameters and conduct a complete review. That work produced a functional specification that met the needs of the Strategic Rail Authority and provided a basis for future work."

His next move was to become a director of Highpoint Rendell, a long established consultancy, where he specialised in finding solutions to technical, organisational and contractual problems.

Then he was asked to become involved in Ascot Racecourse, by invitation from the trustees. It was his job to ensure that the racecourse reopened in 2006.

"That was a huge challenge," he says. "We had to demolish and rebuild the most iconic horse racing venue in the world and realign its historic racing track in a little under 18 months. It was considered by many to be almost impossible – bordering on the heroic, even! But that's what the client wanted to achieve."

The team delivered on time and Ascot's races were held in 2006 as usual. No sooner was that finished when the phone rang yet again.

He was asked to become Director of Construction for the ODA.

"I did some reflecting on what that might mean," says Howard. "I was told it could do a lot for my career and reputation. I said 'yeah, I could lose them both!' But I took the job. It's about managing risk, like all big projects. You have to assess the risks. I weighed it all up – and I took it."

Of course, there have been plenty of challenges since then. To begin with, most importantly, companies from the industry were not queuing up to get involved.

"Everyone was pretty busy, and they weren't that keen about taking on government work under a major media spotlight," Howard explains. "Also, whilst the industry knew some of us, we had no track record as a team. So there wasn't a great appetite for coming to work for the ODA. Over the past two years, we've all worked really hard to demonstrate that we're the sort of customer the industry wants. Now, all the top industry players are absolutely committed at senior level to the successful delivery of a successful games. We're using iconic architects, highly experienced contractors, and we've got some stunning designs."

Howard wants the Olympics to showcase the best of British construction.

"It's about a great sporting event," he says, "but it's also about national pride and capability. We've got an opportunity to create an environment where the industry can give its best and demonstrate what it can do on a world stage. I'm serious about training, about employment, about safety. I want more people to have the opportunities I've had."

'Opportunity' is a word that crops up a lot when talking to Howard. Ever since he was a teenager, it seems, he's seen opportunities rather than setbacks or challenges, even on difficult projects such as Thameslink.

"What motivates me is that I love what I do, and I feel I can make a difference," he says.

"I'm also very lucky that after nearly 40 years of marriage, my wife knows as much about construction as I do! It's great to have a sounding board. That's important. Otherwise, being in charge can be lonely. But I'm not afraid to make a decision. I like to think that people who know me would say I'm firm but fair. I believe wholeheartedly in collaborative working to develop the best solutions. That's my modus operandi."

Datuk Michael Yam

{1953–   }

Member of the Chartered Institute of Building

Michael Yam, one of Malaysia's most successful developers, comes from a family of builders. His father was a contractor for 40 years.

"He got to the top of his trade," says Michael. "He's always been my role model."

However, watching his father taught Michael that the life of a contractor was not an easy one.

"In Malaysia, contractors were too often at the wrong end of the stick," says Michael. "They got bullied. Now we have a more professional industry and things are changing – as they are across the world – but 35 years ago the situation was not good. That's why I was attracted to project management for clients."

Nevertheless, Michael's background ensures that he always has compassion for contractors.

"I urge everyone to honour contracts and give contractors a fair deal," he says. "I saw my dad suffer from the vagaries of the industry – sometimes when times were bad, clients didn't pay. Since construction is all about cash flow, builders need to pay their subcontractor or they go bust. This is ultimately a social problem, because at the bottom of the chain, it's the last one who doesn't get paid – and this often means poor migrant workers. It's a pyramid, and if you view the whole, you see that the ramifications affect more than just business."

Although he didn't want to be a contractor, the young Michael Yam did get insights from his father about running a business. He thinks leadership is probably half nature, half nurture.

"You need an extroverted personality and good communication skills," he says. "I learned on site to be able to talk to anyone – in Cantonese, English or Malay if the need arises. In Malaysia, there often are English contracts and drawings,

built by Malay speakers with Cantonese-speaking supervisors, so this flexibility helps!"

Another early source of leadership training was Michael's scholarship to the Royal Military College, which offered boys an elite education, as well as military discipline.

"I enjoyed it," he says. "There were 100 of us, boys picked from every region and social strata, put in a melting pot. Many of us left to take leadership positions throughout Malaysian society. Without that, I would still be an unpolished diamond, perhaps!"

Entering the army was not a condition of his education, so Michael returned to work for the family business – allowing his father to take the first holiday of his life.

However, Michael had ambitions that would take his talents elsewhere.

"Everyone in those days wanted to go to England," he says. "Even up to the 1970s, almost all the professionals in Malaysia were British educated. There was a premium to being English-trained. Everywhere else was secondary, although now Australia is the destination for many young professionals."

He was taught by Professor Michael Roman at the Polytechnic of Central London, and in {1983}, gained corporate membership of the CIOB.

While studying, Michael worked for a construction company called J Lawson and Co, gaining business insights that would have been hard to get back home in Malaysia. Working on projects such as an old people's home at Wembley and a comprehensive school, he obtained his industrial training as a site engineer.

"The site agent was this unforgettable rough, tough tradesman called Joe Bridle," he says, "and I learned so much from working with him and his men. I was treated as more than an equal by them – they invited me back every holiday, and kept me registered as a labourer so I could earn time and a half on Saturdays! I had fun, but the real revelation was that it gave me something I could never have had in Malaysia. There, I could have been director of a construction firm, but

{1983}

Conoco-Phillips Building in Anchorage, AK is completed. At a height of 90 meters (296 feet) it is the tallest building in the U.S. state of Alaska.

not have known how to construct. In England, I could start from the bottom and learn the nuts and bolts of the business. It didn't matter who you were – if what you did was no good, you were told about it. In Malaysia, where I was the boss's son, no one would say."

Michael then took on an independent project, which he describes as his finishing school, building a mansion in Hampstead.

"The developer was a classmate of mine at management school, and I said I would manage the job," he explains.

> "So that made sure I understood the entire process. I carried hods, drove the forklift, made tea, bought breakfast. Whenever anyone was sick, I did their work. I've always loved the fresh air of construction, seeing things grow up from the ground."

The project was a success, but as the early eighties' recession loomed, he decided it would be judicious to apply for something more permanent. He duly found a job with West Thames Area Health Authority.

"The NHS was a great place to take shelter from the recession," he says. "I was a building officer for what was then the largest psychiatric hospital in the UK. I learned about proper management and maintenance – the skill set which is now known as FM. Much of the job was about dealing with minor works like leaking pipes, but I learned about grounded systems and processes, which has been very useful. As a developer with involvement with new builds, I'm now using exactly the same techniques to manage the trades as those I learned in the NHS. While I was there, they brought in an incentive scheme, so I could manage using performance-based pay. This is important, because it's difficult to motivate people if they get the same amount of money whether they fix 10 leaking taps or 20. That was just coming in then, but it remains a useful tool for managing work on investment properties."

By then, Michael had met and married Cindy and celebrated the birth of Grace, the first of their four children, in 1982. The following year, they returned to Malaysia to make it easier to have a good standard of living on a single salary.

Ron Whitehouse who had a distinguished career with Taylor Woodrow as its international Managing Director, hired Michael for his Malaysian project management consultancy known as PDCS Limited and became his mentor. Here, Michael learned about the importance of measurements and accuracy, and how attention to detail early on could avoid claims and disputes later. After six years with the firm, when he'd become the first local director with a Malaysian company, a major hotel group (Landmark Bhd) offered him the role of group property manager.

This led to work on a couple of significant projects. He was responsible for the renovation of Sungai Wang Plaza, built in the seventies by another Malaysian construction leader, Datuk Dr Bernard Wang. He also had the challenge of transforming the old residence of the British High Commissioner into the 'state guesthouse of Malaysia'.

It was a significant heritage building, and the first guest to stay would be Queen Elizabeth II.

"There were 40 acres of grounds," he remembers, "and it was a nightmare getting the place ready."

Michael's portfolio as project manager stretched to ensuring that the Swiss chef flown in from Korea knew how to prepare Her Majesty's kippers for breakfast, and liaising with the security forces about tight security.

"There was a whole battalion of armed forces round the hotel, in case of a terrorist attack as it was during the height of IRA activities then." he says.

From 1993, he was in charge of both property and logistics for Peremba, an investment holding conglomerate.

"Project management is transferable," he says.

As always in his career, Michael was consolidating the skills he'd learned previously and acquiring new ones. He was overseeing major developments, including Kuala Lumpur's Wangsa Maju township, which has become one of the largest in the city.

In 1996, Ron Whitehouse called to tell him that Country Heights Holdings BhD was looking for a group CEO.

"The business was well known for upmarket housing projects, but was just going into hotels and offices and needed specialised professional help," he says. "It was a step in faith, requiring people skills and financial skills, as well as a technical background."

Working for the company's founder, whom he describes as visionary, Michael found himself learning again.

Mont Kiara Township | Kuala Lumpur, Malaysia

{1998}

Bellagio Hotel and Casino in Las Vegas, designed by Marnell Corrao Associates opens.

{1998}

Chek Lap Kok Airport in Hong Kong, designed by Norman Foster is completed.

{2002}

On July 5th Imperial War Museum North in Manchester, designed by Daniel Libeskind, opens.

{2002}

Falkirk Wheel, a rotating boat lift, near Falkirk, Scotland, is opened by Queen Elizabeth II as part of her Golden Jubilee.

Mont Kiara Township | Kuala Lumpur, Malaysia

"He had the big picture, and he would go into a market on a hunch," says Michael. "That's how we ended up creating the Pecanwood Estate in South Africa. It was just after the end of apartheid, and he could see potential. But to follow this through, I'd have to ensure we understood the law, exchange rates, interest rates... I had to structure the entire project. When he had a hunch, I had to go in and make it work."

By now, Michael was also active within CIOB Malaysia, and also the Real Estate and Housing Developers' Association.

With Michael as CEO, Country Heights quickly began to show record profits. After only a year, he accepted an offer to head up Sunrise PLC, a development company.

"I thought being captain of an Airbus was better than being co-pilot of a 747," he once remarked of this move.

Unfortunately, the Airbus was flying into a turbulent climate – the Asian economic crisis hit just months after his arrival. A mixture of experience and faith saw him and the company through.

"I became a Christian in the early nineties when the kids were young," he says. "I look to the Bible for guidance – it's a way of life. I think that if you espouse a faith, you have to try to walk the talk, but I also believe God is merciful and doesn't expect us to be perfect."

For Michael, therefore, the company's survival can be ascribed to the grace of God.

"There is a saying – man proposes, God disposes," he says. "1997 and {1998} were impossible. It wasn't in any textbook, but we survived. I'm just hoping not to get called to do anything like it again!"

With the cashflow dry, Michael had to talk honestly to his staff: they elected to take a pay cut rather than see redundancies. He also started developing smaller, more affordable condominiums as a 'recession product'. These sold quickly.

In a country where owner-managers are still the norm, Michael is one of the new professional CEOs who have set a benchmark for successful property development. In {2002}, he was both made a Datuk and recognised as Malaysia's CEO of the Year, and he gets a great deal of positive coverage from the Malaysian press.

Although he is now in the process of stepping back from his full time executive role, his diary fills up quickly – as is often the case in these situations.

"My aims were to bring down my golf handicap, spend more time with my family and build my dream house," he says. "But my golf has got worse, my dream house is on the back burner, and the best I can say about family time is that it hasn't actually decreased!"

This may be related to the fact that he's become advisor to five different businesses – one for each day of the week, he says – and has interests in various international investments. He speaks at conferences and people seek his opinion on everything from corporate governance to the credit crunch.

And as if this is not enough to fill his days, he is one of only three locals on the Board of the British dominated British Malaysian Chamber of Commerce, the current Chairman of the Real Estate and Housing Developers' Association and a trustee of the Standard Chartered Bank Charity Trust. Additionally, he was recently appointed by the King to the prestigious position of an Advisor to the City and Mayor of Kuala Lumpur and in 2008 he was also nominated as a pioneer member of the Board of Trustees of the Chartered Institute of Building.

He is pleased with his success: Datuk before 50 and in possession of a chauffer-driven Mercedes before the age of 40. But he does not appear overly attached to status symbols.

"It's part of the culture," he says. "If you are seen in an old, battered car and travel economy, the perception is that you can't be doing very well. You have to be seen to arrive at functions in the manner which the shareholders expect, or you could potentially knock the value of the company. People notice these things."

However, as our interview ends, he is going to pick up a hire car for his stay in the UK and appears to be enjoying the opportunity to be an anonymous person in an Astra. London still offers Michael the chance to take a break from his executive persona, just as it once offered him the freedom to work on site without being seen as the boss's son. Having worked all over the globe and progressed on his own merit, Michael Yam has little left to prove.

{1969}

The British Telecom Tower in Birmingham, West Midlands (England) is completed.

{1969}

John Hancock Center in Chicago, United States is completed.

Millennium Dome | London, United Kingdom

Bernard Ainsworth

{1947–   }

Fellow of the Chartered Institute of Building

It was Bernard Ainsworth's career advisor who suggested he might be best suited to a practical career.

Bernard might otherwise have studied for his A-levels, but his experience from a holiday job in a hospital maintenance department convinced him instead to choose an OND in construction. This was followed by an HND and a sandwich course at Liverpool John Moores University.

As Bernard was finishing the first part of his studies, Laings were looking for promising students to train during their sandwich year in industry. Bernard was offered a place – the start of a working relationship spanning 18 years and some major projects.

He remembers learning very fast during {1969}, his year in industry. He worked as a production controller on the elevated Westway section of the A40, assessing targets for tradesmen – a task that had a direct impact on their salaries.

"Everyone had to buy into what we agreed," he explains, "especially when you remember that the chippies had axe belts!"

It was Bernard's first time away from home, but he felt like the world was his to conquer. He was even happier when the time came to return to college, because Laings offered to support his final year – a move that effectively doubled his grant.

"I was incredibly pleased that they offered it all without my even asking," he says. "Looking back, I see it was sensible, because it guaranteed that I'd return after I finished my degree."

Indeed, after graduating he spent nine months working on the M5, and acquired his first piece of company transport – a 125cc Phantom motorbike.

"You needed it on that job," he says. "It was less intensive than working in London, but there was a lot of ground to cover."

As that project finished, he was given the chance to work in Saudi Arabia. The offer was made on Christmas Eve, and he found himself there in early January.

"That was an interesting experience, but one I would only recommend to young bachelors," he says.

"I worked with eight or nine lads of similar age. We worked hard and had a whale of a time – I think we drove the project manager mad. Looking back, I'm pleased to say I did it, but at the time, I sometimes wondered what I was doing, with nothing to do but work and go to the bar. There were no women there at all. It was a very unbalanced existence. Even so, while inducing premature grey hairs for his project manager, he helped build the royal Saudi airport and worked on several smaller sites all over the country.

"I was sent to the Yemen border for a while," he says. "I got invited to some of the local events, such as weddings. I remember sitting on a blanket sharing a cooked sheep."

On his return to the UK a year later, he found himself posted to less exotic surroundings. A new power station was being constructed on the Isle of Grain, in the Medway Valley of North Kent. It was a heavily industrialised area, and Bernard was one of several hundred workers billeted in a small village near the site.

"I think it was exciting for the village because we brought money into the area," he says. "A new pub was built while we were there! In other ways, it must have been very disruptive.

There was a camp next to village and we also bussed people in. I was there for five years – I even ended up buying a house! The project took so long that abiding connections and friendships were formed."

Just as in his first role, he was given the job of production controller on the project.

"I started with a section, but ended up production controller for the whole thing," he says. "It was my job to manage people and support the managers by giving them information for decision making. It was wonderful career development. I learned to converse, to manage people, to understand figures and to grasp how things fitted together.

He also had a hand in managing industrial relations during a two year laggers' strike.

"We were bussing people through the picket lines. It was difficult," he says. "At the time, fraught industrial relations were a regular part of the working world. But dealing with union stewards taught me a lot about management in tight situations. We were arguing about things that really mattered – the bonus rate was such a large slice of the wage that it nearly overbalanced the wage rate. Discussions got emotive. There was a lot at stake."

{2002}

Beddington, London,
designed by Bill Dunster
is completed.

{2002}

City Hall in London,
designed by Norman
Foster, is opened.

{2002}

Federation Square
in Melbourne, Australia
is completed.

Bernard thinks that despite those days being tough-talking, it's a shame that young managers today don't get the same opportunities to hone their negotiating skills.

"You don't get the same interaction with the labour force because that's all wrapped up by the trade contractors," he says.

As Bernard's time on the Isle of Grain was reaching a close, Laings put in a bid to build a power station in Poland, a project to be led by Bernard's colleague John Armit. Bernard was offered a chance to join the team as a planner. Before he'd even left the country, however, the project taught him an interesting lesson about management.

He was about to get married to his fiancée, Rosie, when the opportunity arose. He rang her to find out what she thought, and she agreed he should go. Bernard might have thought no more of it, had a colleague not taken him to one side.

"Bill Dewis was a hard-bitten, old-school engineer," says Bernard, "but he told me to go home and talk to Rosie properly."

It transpired that Rosie was not totally thrilled about the situation after all. So it was arranged for her to travel to Poland with her new husband and to become the site nurse.

"There have been three or four really important mentors in my life," says Bernard, "and Bill is one of them. You need mentors, and in time you need to become one yourself. That incident taught me always to think about people's family circumstances. You have to find out what people really feel. If you find out what they really want, you can match it to whatever it is that you want to do – and get the best out of them."

On his return from Poland, he was sent to work at Boscombe Down, an RAF airfield.

"I went in as the number two on the project, but the project manager was moved, so I ended up in charge," he says. "We were building shelters for F11s, very much part of the whole Cold War era. That's now history, but I was really surprised to find that English Heritage is listing the site. That's the last thing I'd think of listing myself!"

Following a stint working on the M25, he was given a cluster of projects to oversee.

"Going from single projects to managing an area I found a challenge," he says.

"You learn a lot about having to rely on other people and trusting yourself and the team you've chosen. However, I realised I liked single jobs better, things I could manage hands-on rather than from far away."

Despite this, in the late eighties Bernard won a series of large jobs for Laing and then left them to run without him, following set-up. One job where he got more closely involved, however, was on a major build for Toyota, the Japanese firm.

"They taught me a lot about taking care of other people, about getting decisions made inside a team and about management," says Bernard. "It showed me how you could get people to really buy into a decision."

After an intensive bid process, Laing had won the contract to build a £200 million car plant. Once Toyota had awarded them the contract, they invited Bernard over to Japan. He was taken to visit existing car plants, but they also spent time talking and playing golf.

"They showed me their existing buildings," says Bernard, "but they showed me more of themselves. As it was the first time they had made a major investment outside of Japan, they said they wanted to get to know me, and for me to get to know them. That wouldn't have happened in the UK at that time."

Their ethos has remained an abiding influence on Bernard.

"It changed the way I work," he says.

"My mantra is: it's about the team. Clear the boulders from in front of the team and they'll produce anything."

During the nineties, Bernard became Laings' operations director for the Midlands, the north and Scotland, reporting directly to the MD. More major projects came under his direction, such as the Inland Revenue headquarters in Nottingham, and the Scottish Widows office in Port Hamilton – the heart of Edinburgh's financial district.

Then, the project of the millennium, as it were, came along: to build the Dome at Greenwich, in London. He thought it looked interesting and despite initial difficulties related to funding, public sector backing gave the project the go-ahead.

Bernard was put in charge. He developed a good relationship with the client representative, David Trench (son of Sir Peter Trench) and also learnt a host of new skills – such as talking to the media on a regular basis.

"It was an intensive political football, but we just got on with it," he says. "It was on time and on budget and it taught me how to get the best out of a team. I did a bit in front of the cameras, and I enjoyed it – I knew we were doing well. The last time I visited, it still looked pristine. The media coverage was annoying, but once the press get their knife into a project, there's nothing you can do."

As he points out, the initial poor reviews of the Dome didn't deter six and a half million visits to its 'Millennium Experience' exhibition, and the space is now enjoying a renaissance under the branding of the O2 arena.

As Bernard finished the project, Ray O'Rourke was doing the famous deal to take over Laings. Bernard was ready for a new challenge and one showed up, in an invitation to get involved in the delivery of the {2002} Commonwealth Games, in Manchester. He duly took leave of absence from Laings to accept the role.

"I'd never have thought to go for the job," says Bernard, "but when I was asked if I was interested, I was really pleased to get the opportunity. It took me away totally from construction, and gave me a lot more competencies."

His construction experience stood him in good stead, however, and he forged a strong team. As he points out, any project that requires 1000 taxis is potentially a logistical nightmare, but the Games were a success.

At this juncture, he retired from Laings, but it wasn't in his nature to give up work for long. After a spell managing phase three of the trams project in Manchester, he spent time working for Atkins. He then moved to T4M, working on a major project delivering underground stations.

He's now directing a project that's almost as high-profile as the Dome itself.

"Everyone will look at the Shard," he says. "In five years, it'll be an icon."

As ever, Bernard is focused on creating the best team possible.

"It's my job to make sure it's an integrated team regardless of contract style," he says. "It's a fixed price job, but the team – and that includes everyone from top to bottom – have one mission. This project is every bit as pressurised as the Dome. We're constantly striving to achieve what we've set out to do – so the whole team has to deliver."

"I've always enjoyed it," he says. "Something always challenges me. I'm also very lucky because whatever I've done, my family have always supported me. The fact that my family, especially my wife, have ridden with this – despite the times I spend away from home – is what makes it all possible. In construction, it also helps if you're intelligent and lucky. If you have these things, and you're prepared to work hard, it's a stunning career."

Professor Li Shirong

{1957– }

President (2009–10) of the Chartered Institute of Building

Li Shirong was born in the town of Ya'an in Sichuan province. Her mother was a primary school teacher and her father was a civil engineer, which she cites as the main reason she chose construction as a career.

Her relatively comfortable start in life might well have provided a good base for professional advancement in a different environment, but it was not the case when she was young. When Shirong completed her schooling in 1976, the Cultural Revolution had not quite ended and the 'Down to the Countryside' movement was still marshalling China's educated urban youth into back-breaking labouring jobs on the nation's farms.

Shirong worked on a wheat farm on the banks of a river, and her tasks included carrying fertile mud from the surrounding mountains to mix into the sandy soil. She had to share a room with two other young girls, and had no idea how long her stint in the countryside would last.

"That time has now faded into history," she says, "but for me it was a very significant experience. It was so difficult and challenging and hard, but it left a positive legacy. I talk to friends of the same age, and I know we all cried because the work was so difficult, but it made us stronger. Ever since then, we have treasured our opportunities. Now, we feel we can always overcome difficulties. It made us appreciate life, and want to work hard to achieve as much as we could when opportunities finally arose. We treasure and respect opportunity. After that experience, challenges at work, in universities and in government didn't seem so bad. I never forget that I had a hard life and now it's much better."

Although some young people spent many years on the farms, luckily that did not happen to Shirong. In {1977}, policy winds shifted in her favour as young people were allowed to take examinations to enter into universities for the first time in a decade.

With ten years' worth of candidates, the rivalry was fierce. All the hopefuls, sometimes referred to as China's Lost Generation, were competing to enter university in the same year. Shirong characteristically resolved to be one of the successes, despite the odds. Even though her working days were gruelling, she chose – along with another roommate – to study during the night, working quietly and using a torch, so that the third person in the room could sleep.

Having passed the exam, Shirong was enrolled in the civil engineering programme at Chongqing University. She began her studies in January {1978}, a proud member of the first cohort to enter university under the new policy since the Cultural Revolution began.

After graduating in {1982}, she was selected to be a teacher in the civil engineering department. Alongside this work, she started studying on the Masters programme in Construction Management. For China, this was a new and exciting discipline, and realising that she would need to develop an international perspective too, Shirong started studying English. She was also appointed to sit on the national Steering Committee for Construction Management Education under the Ministry of Construction.

Her career progressed well and she gained a promotion to associate professor as her interest in Construction Management continued to grow.

With the Chinese approach to economics undergoing a seismic shift, the need to understand western, commercial approaches to building increased rapidly.

{1977}
Sainsbury Centre at the University of East Anglia in Norwich, England is completed.

{1978}
Gehry House by Frank Gehry in Santa Monica, California is completed.

{1982}
November 13 – Vietnam Veterans Memorial in Washington, DC, designed by Maya Ying Lin is dedicated.

{1993}
Landmark Tower in Yokohama, Japan is completed.

In {1993}, she won a scholarship for a year's study at Delft University of Technology in the Netherlands. The challenges of another foreign language and separation from her family made it a difficult year, but Shirong never lost sight of the strategic importance of studying in the West.

Before she left China, Shirong had become aware of the work of the Chartered Institute of Building due to her connections in the Ministry. Living closer to the CIOB's headquarters, she became keen to join.

"Going to Holland brought me so many good opportunities," she says. "I started to hear more about the CIOB and I felt there would be a lot of benefits for me if I joined."

Such ambition was not based purely on self-interest, however. Shirong never seems to have differentiated between her own personal development and her desire to assist progress in Chinese education.

"Not only was it a personal opportunity for me," she says, "it was clearly going to be a good link for collaboration as China developed construction management courses."

Only the third Chinese national ever to join the Institute, Shirong's initiative was to benefit the CIOB as much as it did her and her department. The CIOB had already tried forging links with the Chinese Ministry of Construction, but was finding communication (including getting hold of English speakers) very difficult. In that context, a request from an English-speaking university teacher with Ministry-level responsibilities and connections was the answer to the Institute's prayers.

Interviewed for membership at Reading University by Roger Flanagan, another synergy presented itself. The University and the CIOB needed someone to finish a government-funded research project on China, due to the unscheduled departure of the previous incumbent. Despite the prospect of even more years separated from her family, Shirong grasped the opportunity to study for a PhD from one of the UK's best universities in Construction Management.

She also began the work of completing the government's three-year research project within a two-year time span. It was a very difficult period; she not only missed her family, but also suffered from tinnitus, a constant ringing in the ears, due to the pressure of work. However, she managed to write the book on time and submitted her thesis in July {1998} – returning to China having achieved her goals and tinnitus-free.

Promoted to full professor at Chongqing University in her absence, Shirong redoubled her efforts to develop the nascent Construction Management steering committee for the Ministry. She led delegations, made introductions, translated and generally fostered contact between the CIOB and the Ministry of Construction. Her contribution is a major factor in the now significant number of CIOB members in China.

"Looking back, I feel my thinking was very creative at that time," she says. "Contacting the CIOB was just an idea, but I was looking for creative ways of improving the information flow to China."

Even so, she couldn't possibly have known how pivotal the contacts she forged would be, both to her career and to the international development of the CIOB.

Undoubtedly, Shirong feels an emotional connection to the Institute, which both reflects and transcends the impact it has had on her career. When she first made contact, she was isolated and far from home, so geographically closer connections would have been welcome on a very human level. Moreover, by studying a subject so novel to Chinese academia, she was also aware of the need for a professional abode where she could put down roots of a more cerebral nature.

"Just as the Chinese embassy is a kind of home from home when I'm abroad," she says, "the Institute has become my professional home. People are surprised when I say this, but I tell them this is the feeling in my heart."

Renmin People's Square Great Hall of the People | Chongqing Sichuan, China

{2003}

Taipei 101 is topped out to
become the tallest building
in the world.

Shirong's successful career as an academic might have continued until her retirement, but fate was destined to intervene again. In {2003}, the government of Chongqing (a region of some 32 million people) decided that it needed new talent to assist in the modernisation of government and invited Shirong to join the endeavour.

For years, she had been used to the deference normally given to professors by Chinese students. Government would be completely different.

"Now, every day is a challenge," she says. "Every day I'm learning something."

The word 'challenge' appears with increasing frequency as Shirong discusses the adjustment of moving, at the age of 46, from the hierarchical world of academia to the multi-stakeholder environment of government.

As vice mayor of Shapingba District, with a population of 1.1 million, one of Shirong's jobs was to lead the team delivering the Shapingba "university town" – a project uniquely Chinese in scale. In {2003}, they began developing the urban planning and relocating farmers from a 30 sq km tract of land and putting the infrastructure in place. By October 2008, there were 80,000 students and 10 universities there, with an extra five academic institutions planned.

As well as managing this huge and multifaceted programme, Shirong used her knowledge of international standards to push for another crucial improvement in the municipal development process: proper urban planning. In a region endeavouring to find a route to sustainable urbanisation on a massive scale, with 400,000 people moving off their farms each year, it is a crucial issue.

"I kept on telling people it is so important to do urban planning," she recalls. "The district did not want to spend money on it. This caused a lot of problems. I put a lot of energy into changing people's way of thinking."

She also campaigned to get projects into the hands of construction experts, rather than bureaucrats, and proposed the setting up of the Shapingba Public Works Bureau. This comprised a roving, multidisciplinary team whose remit was to advise on all aspects of procurement and delivery.

In 2007, she joined the Chongqing Foreign Trade and Economic Relations Commission, leading a group of four divisions whose job is to attract inward investment.

With foreign investment doubling in the first eight months of 2008, it is clear that Shirong continues to have creative ideas. One of these came to fruition in September 2008, when the UK and Chongqing signed a Memorandum of Understanding giving the Chinese conurbation pilot status as a "sustainable city". This meant it would set up as an attractive destination for hundreds of UK companies wanting to sell innovative 'green' products and services, from light-emitting diodes to urban drainage systems.

"We're going to create sustainable rural and urban areas," says Shirong. "The MOU encompasses eco-villages, eco-towns with clean rivers, clean coal and conservation. There are a number of projects under discussion. Another action plan based on this MOU was signed between the UK and Chongqing governments when Premier Wen Jiabao visited the UK at the beginning of 2009. I really put my energy into this project. It's high-level and the potential is very exciting."

Equally exhilarating is the development of a centre for professional excellence. The brainchild of Shirong and the CIOB, the idea is to ally British 'soft' skills with Chinese technology.

"We have the facilities and administration," says Shirong, "but the UK will provide a training framework to train our people to get qualifications. Several UK professional institutes, including the CIOB, have already started their training in Chongqing. There are several others keen to be involved as well. The UK institutes get members, whilst we get trained people, so it works for everyone."

Clearly, Shirong has successfully overcome the challenge of becoming influential within Chinese government. And now, becoming the first female president of the CIOB, let alone the first Chinese president, brings yet more tests. Shirong is keen to use her term to ensure that international communications, with a particular focus on sustainability, remain high on the CIOB's agenda.

"As the world becomes smaller and smaller," she says, "it's easier to get in contact and share experience about how to make the best use of our resources. The world is challenging – we need communication and collaboration internationally, particularly during the current economic crisis. This will make a lot of people pay more attention to cost and value for money. Therefore, we should develop platforms for members to communicate. For example, if we could take a UK contractor, perhaps working in the Middle East, and put them in touch with a Chinese contractor who can build wonderful buildings, but needs international management experience, they could learn so much from each other. It's not about performing the function of a trade organisation, but supporting the technology to improve the industry. Sustainability should also be a major focus of our discussions."

More than thirty years after labouring in the mud, Li Shirong is still breaking new ground. Modestly, however, she puts the achievement within a much wider context.

"I am just a normal person from China," she says.

"For the Institute to trust me and elect me is a big thing, for both myself and my country. I will never forget my daughter's reaction to the news. She said that this was the most important thing in my life, as it shows that as a professional I got to the top. She told me I must take it. But this post really isn't about me. It's about changes in China. This couldn't have happened thirty years ago, under the planned economy.

Through the open door policy and reform, we changed – to be able to compete in international society. It's an historic achievement."